AF470748

Robert Temple is the author of the best-selling non-fiction book, *The Sirius Mystery*, which appeared in ten languages. Between 1978 and 1980 he co-edited the American magazine *Second Look*. Since then he has contributed to many periodicals, including *New Scientist* and Time-Life's science magazine, *Discover*, concentrating on new developments in science. He has published profiles of many of Britain's most distinguished scientists and is a Fellow of both the Royal Astronomical Society and the British Interplanetary Society. Temple is also a seventeenth-century historian and has contributed to many scholarly journals. He lives with his wife, Olivia, a landscape painter, in the West Country.

Strange Things

A Collection of Modern Scientific Curiosities

ROBERT K. G. TEMPLE

SPHERE BOOKS LIMITED
30–32 Gray's Inn Road, London WC1X 8JL

First published in Great Britain by
Sphere Books Ltd 1983
Copyright © Robert Kyle Grenville Temple 1983

TRADE
MARK

This book is sold subject to the condition that it shall not, by way of trade or otherwise, be lent, re-sold, hired out or otherwise circulated without the publisher's prior consent in any form of binding or cover other than that in which it is published and without a similar condition including this condition being imposed on the subsequent purchaser.

Set in Century Schoolbook

Printed and bound in Great Britain by
Cox & Wyman Ltd, Reading

For Christopher and
Jean Serpell,
friends of many years

ACKNOWLEDGEMENTS

I wish to thank my wife, Olivia, for her encouragement and for suggesting this book in the first place.

I am also grateful for particular help or assistance in the preparation of this book from Bob Chapman, Anthony Goff, Roy Markham and Dr Julie Serpell.

FOREWORD

One of my favourite pastimes is searching for bizarre reports and discoveries at obscure scientific conferences and in periodicals. Some people play tennis, others go fishing or sailing. My sport is hunting Strange Things in the scientific jungle. I don't kill them; I try and bring them back alive.

What is a Strange Thing? Ideally it should be amusing or interesting in itself. But primarily, it is *anomalous*. One dictionary defines this as meaning 'out of keeping with accepted notions of what would naturally be expected'. Because I am convinced that our notions of what should be expected are inadequate, I continually search for the anomalies which show the cracks and chinks in those notions. Fortunately, as a sport, the quest for these Strange Things is delightfully amusing. A really good Strange Thing is funny enough to justify a laborious chase through the underbrush of science.

Some teachers try and tell their pupils that science can be fun. That is an understatement. Science can be a riot. We just have to get our priorities right. It is never any use trying to take anything entirely seriously. All the subjects which cause people to get tense and stiff should be tackled in a more good-humoured manner: religion, sex, politics, one's self. As Professor David Bohm, the quantum physicist, has said to me: 'All thought is play'. People who can't have a good laugh about most things are in bad shape. When you allow yourself to become humourless, you begin to take yourself far too seriously and you become pompous. As soon as that happens, you begin to be unable to spot the chinks and cracks in the accepted notions. And then you become the victim rather than the master of thought. This is the way downhill, for people and for civilisations. The Romans took themselves too seriously, and look what happened to them.

In the book which follows, I have credited the scientists whose work I have reported or drawn upon by name, because I believe scientists should receive more individual credit and attention than they do. They deserve to be applauded for their achievements. But none of these scientists is in any way responsible for what I have said, or should be assumed to have given his authority for anything in this book. I absolve them all of any responsibility whatsoever, especially for the jokes. But I hope they will all be good-humoured enough to realise that I have tried to make their important work accessible and enjoyable for the public, while being as accurate and reliable as I can.

THE BIGGER THEY ARE, THE FEWER

If you were to weigh all the elephants in a certain part of Africa, and then weigh all the mice, the total weight of each would be about the same. The amount of mouse-flesh would equal the amount of elephant-flesh. A biologist has recently discovered what may be the most bizarre law of nature on this planet: in any given animal habitat, the total amount of flesh of any species will be more or less the same as the total amount of flesh of any other local species, regardless of body size.

This also means that the total amount of energy used by any given species is the same as every other species in its neighbourhood: sloths or possums use the same amount of energy as antelopes or gazelles. Any free-ranging, unendangered species therefore is rather like an 'energy-blob' of equal size to the 'energy-blobs' of any other species roaming the Earth. The different species may distribute themselves in small pieces, like mice, or in large pieces, like elephants, but it all amounts to cutting equal cakes into thick or thin slices. With a cheese slicer you may get two hundred slices from a chunk of cheese, whereas with an axe you may only get two or three. But no matter what size the slices are, the total amount is the same size. And with species of animals it would seem that this is also true.

Another way to look at it is to imagine someone given a certain capacity of electric power. He can take his pick how he wants to use it. He may use it to light one thousand Christmas trees if that is what takes his fancy. Or he could run a hundred electric stoves and produce meals for a variety of restaurants. Or he could run a dozen huge ovens and have a chain of bakeries. Or he could use all his electricity to power a small factory. The same amount of power could run anywhere between one and a thousand individual things. The more things, the smaller they have to be.

Hitherto there did not seem to be any reason to think of species bearing any particular similarity to one another. After all, individual creatures look so different that it is difficult to see what a mouse has in common with an elephant. But now it appears that equal amounts of energy have been chopped up into a few large pieces - elephants - and a much larger number of small pieces - mice.

Obviously, this principle only holds true when talking about common local regions. It is no use comparing the amount of flesh of polar bears to the amount of flesh of camels, as camels do not live in the Arctic nor polar bears in the desert. We have to talk about groups of animals in the same areas.

Animal habitats all over the world have been studied to test this idea: America, Mexico, Malaysia, South Africa, Poland, Uganda, Sri Lanka, Canada, Peru, Norway and so on. Also, the studies made so far only concern herbivores - plant-eating animals. Carnivores were not included in the study, but it is suspected that they would show similar patterns. The study was only published in 1981, and its author announced he was continuing and extending his work. He believes he has found a new 'general principle' of nature, and it would seem that he has.

Since it is the amount of flesh which seems to remain constant, and merely the numbers of individual creatures which vary according to creature-size, we should consider just what flesh is. In their book, *Ecoscience*, Ehrlich, Ehrlich and Holdren have an eerie paragraph which repays quoting in full:

> 'All flesh is grass. This simple statement summarises a basic principle of biology that is essential to an understanding of both ecological systems and the world food problem. The basic source of food for all animal populations is green plants – "grass". Human beings and all other animals with which we share this planet obtain the energy and nutrients for growth, development and sustenance by eating plants directly, by eating other animals that have eaten plants, or by eating animals that have eaten animals that have eaten plants, and so forth.'

So if we consider this, we see it is really the grass that is remaining constant from species to species, or the amount of grass taken in by the species and turned into flesh. And this may be a characteristic quantity derived from the size or fertility of the planet Earth itself. Every species seems to have been issued with its book of ration coupons by Mother Earth: so much grass per week, no more, no less.

What is man doing to this arrangement with his population explosion? With his technology he completely overrides the ration coupon system. This goes far beyond the world food problem. It is not just that he will be running short of things to eat; he may be breaking some extraordinary natural principle or law of species-mass which might bring some unknown disastrous consequence which we cannot yet imagine.

One consequence which may come about is that mankind may slip down a notch on what is called the food chain, to compensate for his numbers. It takes 10,000 pounds of grain to produce 1,000 pounds of beef in cattle, but 1,000 pounds of beef can produce only 100 pounds of human flesh. This sequence is what is called a 'food chain'. In moving from one link in the chain to the next, ninety per cent of weight is lost. But by shifting downwards from link three to link two, human beings could eat grain and produce 1,000 pounds of human flesh from 10,000 pounds of grain, bypassing the cattle stage.

No surveys appear to have been done, but the number of vegetarians in advanced societies seems to be increasing drastically. This is largely through choice, and seems most common among younger people. It is possible that this may represent a compensating mechanism by the human species- essentially unconscious- to meet the strains imposed not only on world food supply, but on the system of ration coupons issued to species by Mother Earth. Sliding down a link in the food chain would decrease ten times the amount of 'grass' consumed by human beings and would entitle us to, perhaps, a tenfold increase in numbers of individuals. But although this might bring us closer to a natural balance, it seems fair to suppose we will always outweigh both elephants and mice!

Source: Dr John Damuth; Committee of Evolutionary Biology, University of Chicago.
(The speculations above are not his - his report was limited to announcing his discovery of 'Population density and body size in mammals' in technical form.)

DEAFNESS MAKES YOU PARANOID

We are used to deaf people leaning forward and saying 'What's that?' But now comes evidence that what they may really mean is 'What's that you're saying about *me*?'

It is not people who are born deaf but people who slowly become deaf that we are talking about here. Experimental evidence was published in the middle of 1981 which proved quite conclusively that the onset of deafness in a person brings with it the very serious risk of the development of paranoia.

When people are getting old they often begin to go deaf. One reason why so many of them refuse to acknowledge the fact is that to admit they are going deaf is equated in their minds with admitting they are growing old. And no one likes to face up to that. Often, of course, they are not really aware of what is actually the matter with them. Deafness can develop so slowly that it is not recognised for what it is for some time.

When a person's hearing starts to fail, he often begins to think that people are whispering or lowering their voices. He then assumes that they may be saying things about him which are not very nice, and as the psychologists put it: 'Denial by others that they are whispering may be interpreted by the hard-of-hearing person as a lie since it is so clearly discrepant with observed evidence. Frustration and anger over such injustices may gradually result in a more profound expression of hostility.'

After a while, as his deafness increases, the occasions when people appear to be whispering conspiratorily can no longer be explained in the person's mind other than by interpreting the events as a campaign of persecution. Social relationships begin to deteriorate quite rapidly and seriously. This in turn deprives the person of the necessary social feed-back to correct his false ideas, for isolation is a notorious breeder of fantastic notions. Furthermore, outsiders witnessing the behaviour of a person going deaf often have no idea that deafness is involved and may interpret the person's increasingly strange and alienated behaviour as due to mental illness. Being treated by others as if one is mentally ill can sometimes precipitate the mental illness which was not there before. The person going deaf is therefore caught in a downward spiral.

The final result can often be full-blown paranoia of a very serious kind. A conviction may grow in the person which is strong enough to resist any contrary information, and it may become impossible to prevent the deaf person from developing a total persecution complex. The irony of this is that the paranoia did not begin as anything irrational. As the psychologists say, it is 'an end product of an initially rational search to explain a perceptual discontinuity, in this case, being deaf without knowing it'.

In order to prove this scientifically, a group of three psychologists got together and devised a set of ingenious experiments using hypnosis. They took university student volunteers and selected from them a group of eighteen who were highly hypnotisable, tested as mentally 'normal' (i.e. not already paranoid!), capable of carrying out acts ordered under hypnosis without being aware of it after hypnosis was over (what are called post-hypnotic suggestions), and so on.

The eighteen students were divided into three groups. None of them was told the purpose of the experiments.

One group was instructed under hypnosis to become temporarily deaf, but was told openly that this was to study the effect of deafness in communication. This was the first 'control group'. The second was instructed under hypnosis to scratch an itchy ear after waking, but without being aware of it or the instruction. These two 'control groups' were set up simply to ensure that the results of the main experimental group could not be explained away. Naturally, none of the people in the two 'control groups' became even remotely paranoid. But it is standard scientific procedure to have these 'control groups' to be sure experimental results are actually due to what one is testing, and not just to hypnosis itself, or being given hypnotic commands. Otherwise experiments would always be open to endless, and justifiable, objections.

The main group of subjects were put under hypnosis and told that upon seeing the word 'focus' projected on a screen in front of them after they were awake, they would go deaf. But they were told they would not remember they had been told to go deaf. In order to conceal from them the true purpose of the experiment, the subjects were told they were taking part in an experiment on the effects of hypnotic training on problem solving.

After they awoke from hypnosis, forgetful of all that had transpired when they were in a trance and oblivious of their instructions to become deaf, the subjects engaged in a conversation with two other people which had been perfectly rehearsed and standardised. Scientists were secretly watching and studying them behind two-way mirrors. Slides were shown to the group, the first one saying 'focus' as if it were being used to focus the slides. At this point, the subjects became partially deaf without realising why, or even being aware that they had gone deaf.

In the conversation with the other two people, which had been planned to contain ambiguous remarks, pulling faces, laughter, and other things designed to arouse some suspicion in a person who couldn't hear properly what was being said, the subjects developed 'significant changes in cognitive, emotional, and behavioural functioning'. As noted by secretly observing scientists, the two other people talking to them, and later, the subjects themselves, they became highly paranoid, exhibiting signs of tension, confusion, unsociability, and uncooperativeness. 'Experimental subjects perceived themselves as more irritated, agitated, hostile and unfriendly than control subjects did'.

The subjects were re-hypnotised and thoroughly instructed in what had actually taken place, and cleared of any harmful after-effects. But the very brief and, presumably, harmless experiment was sufficient to show that within a short time a number of perfectly healthy people deprived momentarily of their hearing, without being aware of the cause, began to develop distinct signs of paranoia.

A better understanding of this process might help people going deaf to guard against paranoia, and help those who know them try to prevent its onset by taking special care to speak up and look open and unsuspicious. Taking such care with all elderly people might eliminate many difficult problems both for the elderly people themselves and those who have to care for them or deal with them. It is one clear example of where public knowledge of a problem could cure a lot of human misery just through being better informed.

Sources: Dr Philip G. Zimbardo, Dr Susan M. Andersen; Department of Psychology, Stanford University.

Dr Loren G. Kabat; Health Sciences Center, State University of New York, Stonybrook, New York.

FROGS AND FISH WHICH USE ANTIFREEZE

It is not just cars, but creatures as well, that use anti-freeze. However, don't try to have injections of it when the weather turns cold – it doesn't work with humans. But it is used in hospitals when human sperm is frozen, for without some added antifreeze, sperm would die when put into the deepfreeze. And for smaller and simpler creatures than ourselves, entire bodies can be preserved during freezing, coming to life again when they are thawed out.

There are several magic ingredients of chemicals which prevent the freezing, the most common being called glycerol. But glycerol by itself is not the complete answer to how the process works, indeed the complete process is not yet fully understood. An Antarctic fish named the 'Ice Fish' (*Trematomus*), coping with conditions down near the South Pole, has been studied at some length. Its blood has been found to contain three different, enormously long strings of molecules called glycoproteins. These have the effect of lowering the freezing point of the blood in that, when ice begins to form in the blood, these long proteins latch onto the ice crystals, binding to them very tightly, and stopping them from growing.

Fish which live near the bottom of the sea have different mechanisms for resisting freezing than fish which live near the surface. Fish from the Hebron Fjord in Northern Labrador in Canada have been closely studied. If they are caught and taken to the laboratory and put into water which is as low as -1.73^{0}C, they will not freeze, as long as they are not anywhere near ice. If, however, they are touched with a piece of ice, ice formation rapidly spreads in their bodies and they die almost immediately.

Further studies revealed that the antifreeze was not distributed evenly throughout every molecule of the body, but that it was apparently restricted to the fluid spaces surrounding all the molecules. This is called 'extracellular fluid' and is estimated to constitute about thirty-five per cent of the body fluids – exactly the proportion of body fluids which were found to turn to ice. The glycerol antifreeze was therefore enabling the thirty-five per cent of fluids to freeze, without killing the fish. For the glycerol protects frozen molecules from damage so that when they thaw out they are intact, like the human sperm and human blood in hospitals.

The two different ways of surviving low temperatures are *supercooling* (lowering the temperature of the blood and tissue fluids artificially, without the formation of ice) and *freezing tolerance* (allowing the ice to form through -out the blood and tissues, causing the body to turn rigid and go into suspended animation, and then thawing out and coming alive later on).

Among insects, midge larvae can be frozen solid to a temperature as low as -25^{0}C and still thaw out and come alive again. In fact, this can be done to them over and over again. Below -15^{0}C, ninety per cent of the water in the midge larvae has been found to be frozen solid. Glycerol antifreeze helps in these cases. For instance, in a variety of wasp, the amount of glycerol in the insect body has been found to increase to thirty per cent of all body fluids just before winter, draining away again when spring comes.

In 1982 it was reported that some species of frog had been found which could resist freezing. The Gray Tree frog, the Spring Peeper, and the Wood frog – all land frogs

Frogs & Fish Which Use Anti-Freeze

- can do so. Various water frogs were found not to have the ability.

Experiments were done with Gray Tree frogs and Spring Peepers, where they were frozen for five days. Over a third of their body fluids were found to have become solid ice when frozen to -6°C. They were then thawed out and came alive again quite happily. This can't be carried too far, however, for when the frogs were frozen to -30°C and nearly two-thirds of their fluids were frozen solid, they died.

These frogs are therefore capable of withstanding what is called 'shallow freezing'. A study of the climate where they live, wintering beneath leaf litter in moist uplands, showed that the minimum temperature a frog might have to survive was -7.2°C. This agrees with the 'shallow freezing' of the experiments which they survived. It shows, therefore, that the frogs have developed capacities to survive just the amount of freezing to which they are sometimes exposed in their natural habitat, but no more.

The frogs were studied anatomically and their muscles and bladders were found to contain even amounts of glycerol antifreeze during the autumn and winter. But in mid-May, they had none at all and could not survive if experimentally frozen.

What has not yet been determined is which garages the frogs and fish use, and which corporation is supplying them with all this antifreeze. If this could be discovered, we could all buy stock and make a fortune on dividends. Meanwhile, the next time you refer to someone as a 'cold fish' bear in mind that he may be below freezing point.

Sources: Dr William D. Schmid; Department of Ecology and Behavioural Biology, University of Minnesota.

Professor Knut Schmidt-Nielsen; Department of Zoology, Duke University.

LESBIAN LIZARDS

Whiptail lizards mostly live in desert regions of the western United States, Mexico and South America, and the curious fact is there is not one male Whiptail among them. And yet they lay eggs which hatch out into baby lizards. This phenomenum, whereby the females require no sperm to fertilise their eggs, is called 'parthenogenesis'.

The babies of Whiptail lizards look uncannily like their mothers. In fact, the resemblance is no accident: the babies are actually genetic clones. Without the contributing sperm of a male, these virgin-birth eggs have no variety of genes, so the daughters are just like their mothers.

It may come as something of a surprise to learn that whole species of animals may be 'gay', but this is the case with the Whiptail lizards. They are not at all deterred by the absence of males from carrying their sexual inclinations through to realisation.

A Whiptail gets hold of another Whiptail in her jaws, and pulls her to her. Then she mounts her as if she were a male, riding her for about five minutes. During this time, the mounted female remains still and submissive, taking it all in, perhaps concentrating, or perhaps 'lying back and thinking of England'.

The lizard who is acting the part of the male rubs her genitals against the partner's back and lovingly strokes the neck and back of her partner with her own legs and jaws. She then grabs the partner at the back of the neck, or on the shoulder, with her jaws, and thrusts her genitals against her partner's genitals by placing her tail underneath her partner's tail.

As soon as the genitals are in contact, the mounting lizard shifts its jaw-hold lower down and forms a loop, achieving the 'doughnut copulation' recognised in species of lizards which have both sexes.

There are even more overt male characteristics among the lesbian lizards. Some of them get all blue in the face – literally. Male lizards of related species get blue faces, forearms, and abdomens. And so do some Whiptails. It is a sign of sexual arousal which is meant to excite the submissive partner and make it seem more like a seduction.

What makes some of the lizards submissive partners and others 'mounters'? It seems to be hormones. Before ovulation, when her eggs are at an early stage in their formation, the mounting lizard has lots of male hormones in her bloodstream. But the lizards who lie there receiving all the attention are in a later stage of either ovulation or actual egg-laying, and their bloodstreams are full of female hormones. Then, later in their cycles, they may switch roles. She who is ravished will one day ravish.

Why do the lizards feel that they need to have sexual intercourse with each other? Does nature encourage the fulfilment of some Great Plan, by allowing all this naughtiness? Yes, is the answer. Just as a delicious meal makes you salivate and secrete digestive juices more than a boring meal, and thereby improves your digestion, so with sex and lizards. For studies have shown that Whiptails who have sex with other Whiptails lay eggs much more often and have many more eggs to lay. They are, quite obviously, 'turned on' by it all. Getting excited stimulates the entire egg-laying mechanisms of the body.

Since Whiptails live in close proximity to one another, sometimes half a dozen gals sitting together in the shade of a single mesquite or creosote bush, there is plenty of opportunity to fall in love and get that tingly feeling. No one knows what the incidence is of incest amongst Whiptails, but since they are all either mothers or daughters, it is possible that even more shocking things may go on than those which have just been described. Lizards may be cold-blooded, but they are not incapable of passion!

Source: Dr David Crews; Department of Zoology, University of Texas.

WHAT GOES ON ON THE MOON?

We are told the Moon is a quiet place, with no wind or weather, no seas or rivers, no life - in short, nothing happening. No one could question the truth of this, but...

There is a hitch.

The fact is that things *do* happen on the moon. They are called 'transient lunar phenomena' (TLPs), and there is no certain explanation for them.

The first 'event' on the Moon was reported in 1787 by the astronomer, Sir William Herschel, who, through his telescope, saw something bizarre happening on the dark side of the Moon. He said it was red and sparkling, and resembled glowing charcoal thinly veiled with hot ashes. He wrote an article about it called 'An Account of Three Volcanoes in the Moon'. The trouble is that there seem to be no volcanoes anywhere on the Moon. Our extensive photographs of the Moon show no lava flows which could conceivably explain Herschel's sighting or any of the similar ones reported later on. And, furthermore, the glow of lava is too dim to be visible as far away as Earth. Nor is there thought to be any source of heat for volcanoes on the Moon, anyway.

You may think Herschel had been up too late that night and had eye-strain. Perhaps. But then explain all the sightings which followed his: in the past two hundred years, 1,400 reports have been made of 'strange events' on the lunar surface which classify as TLPs.

That's a lot of eye-strain, and a lot of late nights.

The average duration of a TLP is twenty minutes. No

permanent traces are ever left. Sometimes a TLP may last for a few hours, but if so, it is usually intermittent rather than continuous. TLPs are often reported as twinkling lights. The average size of a TLP light is sixteen kilometres across, with bright spots of perhaps three to five kilometres across within it.

The TLPs are often coloured. Sometimes they are red, sometimes they are blue. They may be white, or they may appear to obscure and cloud over features on the surface. Is the Man in the Moon puffing on a cigarette? Serious suggestions have been made of 'moon smoke'. But where does it come from?

Strange events occur on both the illuminated and the dark side of the Moon, therefore, reflected sunlight cannot explain them all. Even if portions of the dusty soil of the Moon were somehow smoothed out to reflect more light, they could not then 'unsmooth' themselves. They would remain that way permanently. And no TLP has ever been permanent. Therefore, surface reflectivity changes cannot be the answer.

Furthermore, TLPs happen very rarely in the mountains of the Moon, they seem to occur mostly at the edges of maria, or in certain craters. No fewer than 300 TLPs have been reported in the crater Aristarchus. The crater Plato has had seventy-five and the crater Alphonsus has had twenty-five. Areas with rills have also frequently experienced TLPs.

TLPs seem to occur more often when the moon is nearest the Earth. Tidal forces? Or is it just because the Moon is closer and we can see these small events more easily?

Luminescence of Moon materials is ruled out because the only possible source of excitation, the solar wind, is not strong enough.

Thermoluminescence is ruled out because no known material could be strongly luminescent enough to be visible over such a vast distance.

Then what could be going on? The only two explanations which seem to offer some hope of satisfying us are that gas is escaping through fissures, and that lightning is being discharged in that gas from friction of small particles.

The first idea is that clouds of escaping gas and fine particles would form clouds of a mist-like nature – genuine 'moon smoke'. These could obscure features, as has sometimes been reported. On the sunlit side of the Moon, they could also sometimes cause twinkling effects through the sunlight.

The second idea is that on the dark side of the Moon, clouds of moon smoke spark lightning. It has been pointed out that a precedent exists for this. The ancient Roman scientist Pliny observed lightning discharges in the vapours of the erupting volcano of Mount Etna, and in 1944 similar lightning flashes were seen in the clouds over the erupting Vesuvius. When very small particles collide they can generate clouds of positive and negative electricity. But no complete scientific description has ever been worked out for these processes, and so they remain unproved. It is always possible, therefore, that this explanation for TLPs could still go up in smoke.

So where do we stand? Is somebody up there we don't know about? Is the Man in the Moon pulling our legs? Or have 1,400 been 'moonstruck'?

Source: Dr A.A. Mills; Department of Astronomy, University of Leicester, England.

THE NIGHT SHOCKER

Off the coast of California, some pretty shocking things happen at night. The Pacific electric ray goes on the prowl along the reefs and zaps its prey with powerful electric discharges. For this reason, the ray has been named 'The Night Shocker' by the marine biologists who discovered its habits.

The Night Shocker has the typical flat disc of all rays, composed of the chest fins fused with the head. It can glide along without rising or sinking, using its particularly strong tail for propulsion. It looks a bit like a swimming saucer.

How is it that the Night Shockers, who have feeble mouths and do not grab their food in their teeth like normal fish, manage to have such neat table manners? Almost alone among the denizens of the deep, they don't stick their mouths out and take rude gulps at passing fish. Being more fastidious and elegant, they take their disc-flaps in front and to the sides of their mouths, and wrap their food up as in a napkin before eating it. And the way they do this is absolutely stunning.

Yes, they stun their victims. When they come within about a foot of a nice juicy fish, the ray's electrifying personality begins to show itself. The fish offering themselves for dinner go rigid; their jaws gape open, their top fins stand up straight and they quiver helplessly. This facilitates the napkin trick. The electric charge of the ray can be as high as a thousand watts. Since a normal reading lamp has a light bulb of 100 watts, we are talking about quite a jolt. Having wrapped the stunned fish up in its napkin folds, the Night Shocker then has the problem of how to get the food parcel into its mouth without the use of hands. It generally does a forward somersault, which causes the fish to whoosh along towards the mouth. Sometimes side rolls serve the same purpose. Then the fish is gobbled very quickly. Night Shockers who miss a good meal sometimes do somersaults in the water for ten seconds or so – whether out of frustration or fury, we don't know.

The marine biologists wanted to prove that electric discharges were really being used to stun the fish and they attached bait-fish on long poles beside camera flashbulbs. When the Night Shockers came along and ate the fish, they lit the flashbulbs at the same time. All of this must have looked rather like taking photographs of some celebrity forty feet underwater. And, of course, the Night Shocker is a celebrity – the kind best portrayed in a horror movie.

In the daytime, Night Shockers bury themselves in sand at the bottom and lie there waiting for something good to come along, which they then ambush. But at night the hunting instinct takes over, which must give them a real charge.

Midnight swimmers beware! You may get a nasty shock. It is not just *jaws* that are out there. You could step on a live wire.

Source: Dr Richard N. Bray, Dr Mark A. Hixon; Marine Science Institute and Department of Biological Sciences, University of California, Santa Barbara.

EGGS WITH SUN TANS

An egg has a real problem. Being an egg is really frustrating and exasperating. There you are, sitting in a soft, cosy nest, and mother gets hungry so she just takes herself off and leaves you exposed to the view of any Tom, Dick, or Harry who comes along. You can't do anything; you can't even roll over. After all, you haven't hatched yet.

Eggs With Sun Tans

What can an egg do in a predicament like this? How can you hide from predators? There is one way – camouflage. You get laid covered in speckles.

But the trouble with this is that usual pigments, such as humans have in their skins, don't reflect enough light, and as a result the body gets hotter. After all, white reflects light, whereas dark colours absorb light.

So it would seem that there is no way out of the dilemma. If you try and cover yourself with speckles so that you look like the grass or the bushes, you get hotter and run the danger of frying to death. But if you don't cover yourself with some disguise, you will be spotted easily and eaten by some fiend.

Fortunately, there is a solution. There are special egg pigments, used for egg suntans, which have an extraordinary ability to reflect the hottest wavelengths of light (the infra-red wavelengths), with about ninety per cent efficiency. So, using these special pigments, which are quite different from the pigments used by other creatures, including man, the eggs have a merry time playing hide and seek with their predators. With normal pigments using melanin, egg temperatures would rise when exposed to the sun by 3°C, and would be killed in half the time of exposure.

Blue eggs reflect better than brown or red speckled eggs. The trick is used by green plants, which need to absorb sunlight for photosynthesis, but avoid getting burnt. And there is an obscure tree frog which has cottoned on to the trick too. But, basically, it is the eggs who are the geniuses (maybe that's the origin of the term egg-head). And if it weren't for their suntans, they would all be omelettes.

Sources: Dr G.S. Bakken; Indiana State University.

Dr V.C. Vanderbilt; Purdue University.

Dr W.A. Buttemer, Dr W.R. Dawson; University of Michigan.

MOTHS THAT USE PERFUME

Oriental fruit moths obviously frequent only the very best and most expensive perfumeries. For they actually use jasmine scent - real jasmine - and that isn't cheap. No imitation ylang-ylang oil for them, no substitute synthetic chemicals. What is more, they even use a fixative!

The female fruit moth first attracts the male fruit moth by laying a trail of sex hormone which he smells from afar, and wings his way to her with lustful thoughts. Panting with passion, he alights near the female and sticks out little scent-producing organs he has called 'hairpencils'. These emit an irresistable perfume of jasmine and the female staggers up to him, butts his stomach with her head, and they copulate.

The scientists who worked all this out and analysed the perfume, killed 5,000 male fruit moths and extracted the perfume from them in order to get enough of it to submit it to chemical analysis. It was described by them as having 'a pleasant herbal odour', which, apparently, the moths themselves have if sniffed by a human when they are flirting with one another.

The moth perfume contains chemical compounds identical to those found in the jasmine flower. No one is certain where the moths get this - do they make it themselves, or is it absorbed from their usual diet of small green apples, or do they make furtive visits to jasmine plants to stock up on it? Most varieties of jasmine are native to the Orient.

Jasmine perfume manufacture is a difficult enough process for humans, much less for moths. Even the very best de luxe jasmine perfumes would normally not contain more than ten per cent of real jasmine scent. It is so powerful that even one per cent added to a cheap imitation will result in a vastly superior product compared with an imitation without any at all. The secret of the scent is an elusive chemical, called a 'ketone', known as jasmone. There is apparently no known way to synthesize this chemical in the perfume industry. How, then, do the Oriental fruit moths do it? This is a subject for industrial research! And they not only have true jasmine scent, they combine it with a cinnamate, copying some of the best perfumeries. For it is traditional to add cinnamic alcohol or cinnamyl acetate to jasmine perfumes in order to 'fix' the odours and improve the strength and persistence of the aroma.

Although it is the males who use the perfume among the moths, the rationale behind its use is still the same as it is for people: sexual allurement.

Source: Dr Thomas C. Baker, Dr Ritsuo Nishida, Dr Wendell L. Roelofs; New York State Agricultural Experiment Station, Geneva, New York.

BRAINS THAT SHRIVEL UP AND GROW BACK

The next time you are tempted to say to your canary: 'You bird-brain!' stop and think what that might mean. The poor little fellow may indeed be rather light-headed in the winter, when he isn't singing and you are cross with him. But there is a reason for it. He has literally lost some of his wits.

Canaries lose twenty per cent of their brains in the winter, but it all grows back again in the spring.

They don't lose their heads entirely, just a fifth of what's in them. The two parts of the brain which shrivel up are those used for singing, and are called the *pars caudale* (ninety-nine per cent of which shrivels away) and the *nucleus robustus archistriatalis*, which turns out not to be very robustus at all, because seventy-six per cent of it shrivels up.

The zoologist who discovered this compares the renewal of these parts of the brain every year with the way trees grow leaves in the spring and shed them in the fall, and calls it a rejuvenation process. Apparently, what happens is that the structures in the brain which enable the canary to have a song repertoire just disappear, and with it, presumably, the song repertoire. In the spring, the twenty per cent of the brain grows back afresh, forming a new network of nerve connections together with the formation of a whole new song repertoire. That is why canaries sing different songs every year.

The spiny part of the brain cells which form the structure are called dendrites. As these grow, nerve connections form along and between them. But if they shrivel away and then regrow, they can do so in different shapes. This is carrying fresh thinking to the limit! We can look at it like this: in Year One, the canary grows a set of dendrites and nerve connections whereby it 'knows' the rules of backgammon, but it can't play chess. Then this all shrivels away, and the canary can't play any games at all for the winter. In the spring of Year Two, it grows a whole new set of dendrites and nerve junctions, and this year it 'knows' the rules of chess, but has forgotten entirely how to play backgammon. And so on year in, year out.

The zoologist actually engages in some provocative speculations based on his discoveries, which are worth quoting: 'The shrinkage of brain nuclei in adulthood, resulting from a loss of dendritic processes, may be likened to a rejuvenating process that reduces the size of a network to an earlier developmental age. Of course, such a process can be labelled rejuvenation only if it is followed by a new wave of dendritic proliferation and "nerve connection" formation. If rejuvenation of brain circuitry ever becomes possible in humans, being able to induce a "shrivelling of existing networks" may be found to be the indispensable first step, to be followed by their regrowth. We may now have an animal model for this kind of phenomenon.'

(In the above quote, I have replaced some technical words with ones more easily understood and have marked them with quotation marks.)

Not all birds are 'bird-brained' in the way canaries are. Zebra finches, which never change their song patterns, have been found not to shed parts of their brain and grow new ones. Therefore, the shedding of brain structures and growth of new ones really does seem to be connected with developing variety and novelty.

One could carry the zoologist's speculations further and say that the ultimate means of shedding the old brain structure is death. For those who believe in reincarnation, this offers a splendid model for what may happen: not only is a tired-out body dispensed with, but a brain structure which is beginning to become an impediment to novel and creative thinking is jettisoned, to be replaced by an entirely new circuitry and wiring in the novel growth of the brain of, first, the foetus, and then the infant, child, youth, and finally, adult. Every time one is born, one learns a new 'song'. General characteristics, such as personality, tendencies, interests, views, and attitudes, need not be jettisoned with the old dendrites, but might somehow be carried over outside the consciousness, deep inside the self which – if you believe in this sort of thing – could survive the transition from old body and brain, to new body and brain. After all, it is the 'song' which is connected with dendrite-structure, not the impulse to sing. Therefore, don't lose heart: if you lose at chess all the time in this life, you may be a grandmaster next time round. And if you always wanted to be a successful pop singer and didn't make it, your song repertoire may be better next time.

Source: Dr Fernando Nottebohm; Field Research Center, Rockefeller University, Millbrook, New York.

BABIES LIKE MOTHER'S VOICE

Mothers have long suspected that their babies know their voices, and preferred hearing them speak to listening to nurses and nannies. Now the proof has been established.

Psychologists took twenty-six babies and carried out experiments with them within about twenty-four hours of birth, and in all cases before the babies were three days old. They gave them nipples to suck which did not give any milk, but which activated different tape recordings. The mothers of the babies were tape-recorded shortly after they had given birth (to make sure their voices had not changed) reading Dr Seuss's *To Think That I Saw It on Mulberry Street*, which must be something babies are interested in, for it really seems to have taken their fancy.

Then the babies were tested. They were coaxed into alertness and had earphones placed over their ears. They were allowed to get used to this for a while and when they were quite comfortable, they were given nipples to suck, which were connected to tape recording equipment. The nurses giving the babies the nipples could not cue or prompt them in any way because they were ignorant of what was being done, could not hear the tapes themselves, and could not be seen by the babies.

The babies were then given a couple more minutes to relax and then their sucking on the nipple was recorded in order to study the sucking patterns, which consist of bursts of sucking, pauses, and so forth. A burst of sucking was defined as a series of individual sucks separated from one another by less than two seconds. Having established the burst patterns, sound recordings of different mothers reading the story were played at the beginning of each separate burst of sucking. By ceasing to suck and demonstrating displeasure, the babies were able to shut off the women's voices which they didn't want, until they found the voices of their own mothers.

For the first day, the babies weren't very good at this. But by the second and third days, the babies became quite expert at obtaining their mother's voice in preference to the other voices. The statistical findings were that the babies obtained their mother's voice twenty-four per cent more often than other voices, demonstrating clearly that they were choosing the mother's voice.

Some of the babies involved had barely had any contact with their mothers. At the very most, they had been with their mothers for twelve hours between birth and the commencement of the testing, in a group nursery during set feeding times. Therefore, the experiments proved that babies, almost from birth, can differentiate between human voices, prefer human voices to other sounds, and above all have a need to hear their mother's voice.

It is thought that babies have a highly developed sense of hearing even while in the womb some time before birth. Familiarity with their mother's voice is believed to begin then. During the last three months in the womb, hearing is improving all the time.

What experiments prove is that talking to her baby is necessary and desirable for a mother, for the fact that babies can't 'understand' is not the point. The baby needs the voice of the mother and wants it repeated again and again. Mothers who coo over their babies are therefore doing the right thing.

It all goes to show that mothers have a lot of power over

their offspring: all they have to do is speak, and their baby's heart strings twang.

Source: Dr Anthony J. DeCasper, Dr William P. Fifer; Department of Psychology, University of North Carolina, Greensboro, North Carolina.

SNAKES WITH HINGED TEETH

Snakes are great at biting, as we all know. But sometimes one has to do more in life than just bite. There are times when a snake, no matter how splendid its fangs, has to think about something else for a change. Such as, how to hang onto a good meal.

There are five types of snakes in Africa, Indonesia, and Malaysia, which have developed moveable parts. They have hinged teeth which swing back and then pop into place again automatically.

The snakes in question, most of which are known as colubrid snakes, feed on 'hard-bodied prey'. The main element in their diet seems to be skinks. Skinks are lizards with hard scales and roughly cylindrical bodies. Not only do they struggle quite a lot when being eaten, but their thick scales make them very difficult to chew. What can the snakes do to hang on to these skinks? Eating skinks with normal teeth would result in the breakage of quite a lot of teeth. It would be as if human beings had fist-fights at every meal, which would have no advantage to anyone, except perhaps dentists.

The snakes developed teeth which were rather flat and curved at the front and which were connected to fibrous tissues acting as hinges. The teeth easily fold back when force is exerted on them from the front, but when force is exerted on them from behind, they lock into place as rigid, downward pointing barriers. This makes it easy for prey to get into the mouth, but difficult to get out. In fact, the hinged teeth operate rather like a ratchet on the skinks: as the scaly skinks slide into the mouth, the teeth swing back and then lock into place on every successive hard scale. This gives the snake a succession of iron grips on the skink without the need to pierce the hard scales with a bite. The forty or fifty small hinged teeth in the jaws of these snakes constitute a gliding surface over which prey may be drawn, as well as a ratchet mechanism for guarding against its escape.

If you are ever in the Malaysian jungle and hear the sound of a creaking door, it may just be one of these snakes with some hinged teeth that need oiling.

Source: Dr Alan H. Savitzky; Division of Reptiles and Amphibians, Smithsonian Institution.

HOW TO BECOME A MEMORY EXPERT

We have two different kinds of memory: short-term memory and long-term memory. Some psychologists decided to try and test whether it is possible to expand the capacities of short-term memory and found a student who was willing to engage in experiments for an hour a day, three to five days a week, for a year and a half. The results were astonishing.

Although they didn't manage to expand the student's short-term memory, the experiments, involving 230 hours of practice in a laboratory, resulted in the student equalling the feats of some of the most famous memory experts in history. The student is described as being an undergraduate, with average memory abilities and average intelligence. He had never had any unusual memory abilities before. But he was able to evolve a system of remembering numbers - using his long-term memory - which implies that any reasonably intelligent person who sets his mind to it, and works hard enough, can become a memory expert after a year or two. The experiment led the psychologists to make the extraordinary statement about human beings that, 'there is seemingly no limit to improvement in memory skill with practice'.

How, then, did the student do it? He was presented with a random series of numbers, at the rate of one number per second, and asked to remember them in sequence. Every time he was successful an additional number was added to the next sequence. At the beginning, he was only able to remember seven numbers in succession. But after twenty months, he could remember seventy-nine numbers in succession!

The student at first tried to retain the numbers in his short-term memory, which we use for everyday things, and which psychologists call 'the rehearsal area', in which we can keep three to six items of information for immediate use. But this short-term memory was completely inadequate to retain long series of numbers. So, the student began to develop a system for remembering, and it was this system which enabled him to accomplish such incredible feats.

The student was a very successful long-distance runner, competing all over the eastern United States. With his avid interest in running, he obviously had a very great knowledge of record running times for a variety of different distances, from the half-mile race to the marathon race – eleven different categories in all. Because of his familiarity with these numbers, he was able to take random series of three or four numbers at a time and view them not just as meaningless numbers, but as running times. For instance, if he were presented with the four random numbers, three, four, nine, two, he would think of them as, 'three minutes and forty-nine point two seconds', near the record for running a mile. The numbers which didn't fit as running times he would interpret instead as ages (eighty-nine point three years old, for instance), or as dates (1944, near the end of World War II). He found that about ninety per cent of all the random series of numbers could be broken down into either running times or ages. Therefore, instead of lengthy series of meaningless numbers, the student converted them into a series of joined running times or ages. One thing this accomplished was to reduce the number of things he had to remember. For he might have to remember fifteen running times instead of sixty separate numbers. It is obviously easier to remember fifteen things rather than sixty.

But reducing is not the whole answer. After all, since he was eventually able to recall seventy-nine numbers in succession - even though they were broken down into twenty-eight running times, we are still faced with explaining how he could improve his performance from seven to twenty-eight, in terms of numbers of things remembered.

A closer study of how the student had operated his memory system revealed that the running times were ordered hierarchically. He used a pattern like this for his record seventy-nine numbers: two four-digit groups followed by three-digit groups, and so on. He alternated between running times with and without decimal points. This gave him a structure. If, say, he remembered 'three minutes and fifty-two seconds', he would have a decimal point and contain four numbers instead of three: 'four minutes, twenty-nine point six seconds', for instance. The rule of alternating between running times with and without decimal points dramatically improved his performance.

The student's system was totally ruined one day when he was presented with a series of letters rather than numbers. For, not being able to use his system of running times (since you can't run a mile in DFG point H seconds), he dropped right back to the beginning and could only remember up to six letters.

The key to feats of memory is therefore the creation of a system of coding things you are trying to remember with familiar things which you already know. You should break the number of things down into sets, such as this student's running times, and then order those sets into layered structures which have further patterns. If you are prepared to work hard enough at your system, and if you try it out often enough to master it, you can become a memory expert, whether you have demonstrated any particular memory abilities before or not. In fact, the working out of the system, which may require fair intelligence, can be done for you by someone else, so that you don't even need to be capable of that, and you will only need to operate the system once it has been developed.

In the ancient world, this was known as the science of mnemonics and was used by orators like Cicero and by poets who recited epics. We are now recovering some of their lost tricks of the trade by psychological experiments.

Source: Dr K. Anders Ericsson, Dr William G. Chase, and Dr Steve Faloon; Department of Psychology, Carnegie-Mellon University, Pittsburgh.

ARE YOU AN 'INVERTED WRITER'?

Some people form their hands like hooks when they write, with the hand above the line of writing and the pen pointing downwards towards the body. This is called 'inverted writing posture' and is the opposite of the more usual posture, where the hand is below the line of writing, with the pen pointing towards the top of the page. Inverted writers always look as if they are hiding some secret and want to stop anybody looking over their shoulders.

Well, inverted writers *do* have a secret. They have abnormal wiring in the visual circuits of their brains. Adopting the contorted pose of inverted writing is their way of compensating for this difficulty. If someone who adopts the inverted writing posture is forced to cease using it and write like 'normal' people, the person so corrected would probably find it extremly difficult, or impossible, to continue at normal levels of study and writing. Therefore, if teachers discipline such people they are doing them very serious damage indeed, not only in terms of learning and education, but also in terms of impairing their neurological and psychological functioning.

What is it that inverted writers are compensating for? Two Canadian psychologists have done elaborate studies of the problem and in the process they managed to improve and correct some previous theories. They got a group of volunteers, both inverted and non-inverted writers, to do some experiments. They had an equal number of right-handed and left-handed people as far as that was possible, but as right-handed inverted writers are extremely rare, they could only find one girl for that category. The other three categories, right-handed non-inverted, left-handed inverted, and left-handed non-inverted, were all equal numbers. The intention was to try and make the studies of inverted writing separate from handedness.

It was interesting that when they tested all the people for sensory perception of touch and sound, they found absolutely no differences between them. But as soon as they tested them for visual perception, dramatic differences appeared, and it became clear that the inverted writers had abnormalities in their visual perception and 'wiring' in the brain.

Unfortunately, the abnormalities, although proven, are not yet fully understood. So although we know that they are there, we aren't sure what they are. The human brain – or the cerebrum, at any rate – is divided into two halves, known as hemispheres. In normal people the visual field is divided into two: if you look at something, everything to the left of it is seen by the right side of the brain, or the right hemisphere, and everything to the right of it is seen by the left side of the brain, or left hemisphere. Thus, there is a criss-crossing in the wiring connections.

What the Canadian psychologists seem to suggest, though proof would be needed from further experiments, is that inverted writers have abnormalities in the criss-crossing of visual signals, such that visual signals from the same side seem to be coming through to the left hemisphere from left-handed people. (They don't deal with the problem of the one right-handed inverted writer they found). One of the most serious problems for inverted writers in our culture is that we write from left to right. This is apparently what is so difficult for inverted writers: they can't 'see' it that way, and need to look down from above and reverse it, as in a mirror. The psychologists

Are You An Inverted Writer?

say: 'In these studies we noted that inverting the hand makes it easier for individuals with a non-inverted writing style to write English upside down so that the text reads from right to left'. In fact, they note that inverted writing is found in only ten per cent of left-handed Jews who write Hebrew - which is from right to left, compared with an incidence of fifty per cent among left-handed people of other denominations in America! It would be very interesting to do studies to discover whether a substantial proportion of inverted writers are of Jewish descent and therefore have countless generations of ancestors who wrote and read from right to left, possibly resulting in some inherited characteristic whereby the visual wiring in the brain developed to accommodate the reading and writing habits.

Studies of Chinese people would also be interesting for a variety of reasons: Chinese descends in columns, and the columns are read from right to left. But also, the Chinese do not traditionally write with pens (though of course many have adopted pens today), and in traditional Chinese calligraphy the position of the brush is rigidly prescribed. Is there such a thing as 'inverted calligraphy'? Has there ever been a Chinese who used his arm and hand in a hook posture to write Chinese characters? The questions and possibilities for future research, if all these lines were to be pursued, appears endless.

However, there seems no reason why we should not be able to know within a very few years precisely what the abnormality of visual wiring is which produces the sort of inverted writing among left-handers in our culture. Right-handers may have to wait longer for the answer.

Source: Dr Morris Moscovitch, Dr Lee C. Smith; Center for Research in Human Development and Department of Psychology, Erindale College, University of Toronto, Mississauga, Toronto.

LIVING CLOCKS IN TEST TUBES

'What time is it?' If you asked somebody that and, instead of looking at his wristwatch he said, 'I don't know, just let me look in the test tube', you might have justifiable cause for surprise. (Not alarm, we're not talking about alarm clocks.)

However, two zoologists and a biochemist from Texas have had the opportunity to tell the time by this method. If this sort of thing continues, test tube clocks may become the latest fashion, and quite replace Mickey Mouse watches altogether on the market. But the only immediate commercial threat is to the egg industry. For, in order to have a clock in a test tube, you first need to have a baby chick. Not only does that mean you will want more eggs to hatch, it also means fewer eggs for breakfast.

The chicken has an internal clock which runs for about twenty-four hours each cycle. It is called a circadian clock. Circadian comes from the Latin *circa diem*, 'about a day'. A lot is known about these clocks which occur in various animals, including man. But what was not known before is that you can take the clocks out of the animals, put them in test tubes, and they will continue to tick over!

In the chicken, the internal clock is contained within a very important gland in the head called the pineal gland. We have pineal glands too, and it may become trendy to leave your pineal gland to science when you die, instead of your eyes, if this research goes on. In fact, mystics maintain that the pineal gland is the third eye, and there are many strange traditions about it. In humans, it is beneath the forehead, and there are cases known of people going mad and maintaining that their pineal glands, i.e. their third eyes, were sticking out of their foreheads on stalks and looking around!

With the chickens, the pineal glands were taken out and the Texan scientists lost no time in attempting to keep the glands alive. Successfully, as it turned out.

The glands were put into incubators inside culture flasks. They were soaked in nutriments which 'fed' them, and were tilted every thirty seconds so that they could be 'gassed' by exposure to a mixture of ninety-five per cent oxygen and five per cent carbon dioxide. These fed and gassed glands survived quite happily for some days.

Since all their bodily needs were taken care of, the glands did not die but instead carried on with their work. And this meant that they continued to operate as twenty-four hour clocks and to produce various hormones and so forth at regular and precise intervals.

Experiments done on the first batch of glands found that they were still at it after two days, so another batch of glands was prepared to see how long they would continue, and they were found to be happily ticking away and secreting their hormones on the third and fourth days as well. The scientists concluded that the chicken's pineal gland, 'is a self-sustained circadian oscillator', or a 'circadian clock'. What they suggest should happen next is that someone try to work out its molecular mechanism, so that we can know how it manages to do all this. After all, how can a gland keep time anyway? Nobody knows. But not all glands are so co-operative. Some chicks just won't give you the time of day.

Source: Dr Charles A. Kasal, Dr Michael Menaker;

Department of Zoology, University of Texas at Austin.

Dr J. Regino Perez-Polo; Division of Biochemistry and Human Genetics, University of Texas Medical Branch, Galveston.

SEX CHANGE UNDER WATER

We would find it very disturbing at a co-ed boarding school if, every time a boy was expelled, one of the girls in the school turned into a boy to take his place. We would find it even more disturbing if she did this underwater, in, say, the school swimming pool.

Something of this kind does happen. But it happens with fish. The fish are called *Anthias squamipinnis* and they live in the Pacific Ocean, off the coast of the Philippines.

These fish live in large social groups where females outnumber males. But although they may be far more numerous, the females are highly sensitive about their males, and if one of the males disappears or is taken away, the females are so disturbed that drastic things begin to happen.

A marine scientist decided to investigate. He went out to the Philippines and found neighbouring groups of these fish, marking their areas with buoys. He swam underwater with them and counted the number of males and the number of females. These groups of fish stay more or less where they are and stick together as social units. So it was possible to go back and visit them day after day, simply by spotting the buoys from a boat.

Then the marine biologist began to cause social disruption by taking a male fish out of each group and studying the remaining fish to see what they would do about it. The removal of the male fish rapidly caused some of the female fish to start to change sex! Sex change in these types of fish had previously been closely studied, and it was known that this process involved highly precise colour changes in five separate regions of the body, and was quite unmistakeable. Sex glands, when examined, proved the point, but males who had previously been females had tissue traces of previous ovaries in their testes.

What was most remarkable of all was that there was a one-to-one sex change act. For every male who was removed, a female changed sex. In fact, fifty-eight males were removed in all, and in response, fifty-eight females changed sex to replace them. When several males were removed from a group at once, only one female at a time changed sex, separated from the next in line by about a day and a half. They never got the numbers wrong, however. It is not known how all this was arranged and computed; why didn't they all change sex at once when more than one male was taken away? The scientist speculates that perhaps there are complex hierarchies of female fish, with the highest changing sex first, before the next down the ladder dares to follow suit.

In the different groups, the females had outnumbered the males by thirteen to two, 294 to fifty, and so on. Therefore, you wouldn't think they would be so fussy about numbers that an exact one-to-one replacement was considered absolutely essential. And yet it was. If the females were so worried about not having enough men around, why didn't some of them just change sex for the hell of it? Why did they have to wait until a male was taken away first? We don't know. But it is a bit fishy.

Source: Dr Douglas Y. Shapiro; Department of Marine Sciences, University of Puerto Rico.

WHO SAYS YOU NEED TO HAVE BRAINS?

If you feel you can't compete with the more intelligent people you know, don't worry. Brains aren't really necessary. You can live a perfectly normal, happy life without them.

This isn't a joke. It's an alarming truth.

A British paediatrician who has made a particular study of hydrocephalic children ('water on the brain' some people call it, a disease which usually involves an enlarged head), has found a student at a northern university who appears perfectly normal but who has almost no brain at all. The doctor says of the student that he has an IQ of 126, has gained a first-class honours degree in mathematics, and is socially completely normal. And yet the boy has virtually no brain.

Inside the brain are four cavities known as ventricles, which are filled with cerebro-spinal fluid and which protect the brain against shock if, for instance, we are hit on the head. A normal human being has somewhat less than half a pint of this fluid inside his spine and brain. About three times a day it is recycled and impurities are cleaned out.

But in hydrocephalic people, the cerebro-spinal fluid leaks or is disturbed in its circulation, and this results in the build-up of fluid pressure. The pressure then causes the ventricles to expand, pressing and stretching the brain against the sides of the head. In young children, whose skulls are soft, the result is that the head is enlarged and swollen.

The student with 'no brain' was referred by his doctor to a paediatrician out of interest, because he had a slightly larger than normal skull, and the doctor thought this might prove interesting at the time. The paediatrician was doing brain scans of people with hydrocephalic conditions in order to study their ventricles and brains through X-ray-type photos taken through the skull.

Imagine the expert's surprise when he did a brain scan of the student and discovered that instead of the normal four point five centimetres thickness of brain between the sides of the head and the edge of the interior ventricle, he found the head almost entirely filled with fluid and only a narrow layer of brain reaching around the edge of the skull, measuring less than one millimetre thick!!

How can a person with a brain only one millimetre thick (less than a fifth of an inch) have an IQ of 126, get a first-class degree in mathematics, and have a perfectly normal social life?

Clearly, the idea that the brain has a lot of excess capacity is proved by this and it all goes to show you don't need to have a lot of brains. This student was an extreme case, but a lot of people have heads ninety-five per cent filled with fluid. To have a head seventy to ninety per cent filled with fluid is far more common. And to have your head fifty to seventy per cent filled with fluid is considered so unalarming that such people are said to have only 'minimally enlarged ventricles'.

Apparently, when the head is so full of fluid, much of the brain is stretched and pressed against the skull, retaining much of its functioning capacity despite the spatial distortion. The brain partly consists of fibres, and they just stretch but continue working. Operations can be done to drain away the excess fluid in brains of this kind, after which the fibres and bits of brain subside back into normal position again. In fact, children with hydrocephalic conditions, if properly operated upon, can in most cases be cured and lead pretty well normal lives, as if they never had a problem. (Unfortunately, this is not as widely appreciated as it should be.)

The brain is composed of 'white matter' and 'grey matter'. It is the 'white matter' which suffers the greater destruction, much of it permanent, from 'water on the brain'. Grey matter is less damaged and makes a better recovery. Also, cases which are slow in developing give the brain time to do some rewiring and compensating, to minimise the harmful effects of hydrocephalic conditions.

Perhaps one of the most interesting questions raised by all this is the role of the human cerebral cortex, which is assumed to be the main portion of the human brain, and absolutely irreplaceable. Can it be that the smaller cerebellum, a brain-within-a-brain, lying underneath the large cortex, takes on much of the cortex's functioning for it, and enables so many people who have 'no brains' to function as if they had? Is the cerebellum essentially another brain, a spare which we can switch over to when we need it? There are more questions raised than there are answers given. What we do know for sure is that we know far less than we thought we knew. But that's nothing new.

Source: Professor John Lorber; University of Sheffield, England.

PRETTY POISON

One of the basic human desires is to be loved. But snakes are different. They seem to prefer to be hated.

The deadliest snake in America is the coral snake. It can be found, in all its varieties, throughout South and Central America, Mexico, Florida, and the south-western deserts of the United States. It is noted for its very brilliant colours. Usually it is seen as a coral-like creature with pink or red skin, circled with bright yellow and black rings. But what many people do not know is that coral snakes can look very different. So watch out where you step! Some coral snakes are entirely black with faint speckly white rings, whilst others are mostly red, with black spots.

Coral snakes are deadly. Their poison works directly on the nervous system. A bite from one of these pretty little things and you are probably dead in no time.

You would think that nobody would want to be confused with such a deadly and hated creature, wouldn't you? Other snakes, adorable little creatures, with no poison, who wouldn't harm a - well, we can't say fly exactly, as they like eating all sorts of things - but, cosy little fellows with no ill intentions who are not social embarrassments, wouldn't want to be mistaken for a coral snake. Would they?

The answer is *yes*! They will seemingly go to any lengths to be thought to be horrid, vile, detestable, bad-mannered coral snakes. What they do is this: they develop the appearance of coral snakes, down to the last detail.

They are coral snake mimics. Hundreds, even thousands, of snakes have been studied all over the Americas. The most amazing examples of mimicry exist which seem to defy physical possibility. Wherever the coral snakes vary in their own colouration and appearance, the mimics vary almost identically themselves. The look-alikes are quite astonishing. And even the rare black coral snakes, with white speckly bands, have exact mimics which are black with white speckly bands.

Why do the non-poisonous, or at least, less deadly, snakes want to be hated? Because then they will be left alone. Predators are afraid of them, and won't go near them. It's a case of the prettiest poison lasting the longest.

Sources: Dr Harry W. Greene; Museum of Vertebrate Zoology, University of California, Berkeley.

Dr Roy W. McDiarmid; Curator, U.S. Fish and Wildlife Service, National Museum of Natural History, Washington, D.C.

SPONGES ARE PEOPLE TOO

A sponge has feelings. So next time one takes a bath with you, careful how you squeeze it.

Of course, these days shops sell pieces of plastic and call them 'sponges', they get soggy and the water in them quickly goes cold, and many people do not even know that there are such things as real sponges which are picked up off the sea bed. Real sponges have become luxury items, so that one large enough to hold in both hands would cost perhaps twenty-five times the price of this book. (Which shows how unprofitable it is to be a writer, and one should be a deep sea-diver instead.)

For those people who do know that real sponges exist, and what they are like, it may come as a surprise to learn that sponges are highly sensitive. It is true that they tend to just sit around and cannot change shape. In fact, the biologists who discovered their sensitivity had to admit that sponges show little behaviour in the usual sense. But we all know that the motionless wall-flowers at the dance are the girls who are most easily hurt. Just imagine if you were a sponge and couldn't show your emotions clearly, but were just rigid with shyness. You would imagine all sorts of slights too, perhaps even think the sea was against you!

Sponges are always nervously 'catching their breath'. It is not the air which they are breathing, it's seawater, but it amounts to the same thing. Three biologists decided to study the secrets of the sponge's 'breathing' and so they set up flow-meters beside some sponges to study their exhaled water.

What became apparent very quickly was that if there was any jarring or vibration, the sponges stopped 'breathing' and 'held their breath' nervously. But this seemed difficult to believe, for sponges are unable to contract themselves, and therefore, how were they able to hold their breaths when there were no muscles or muscle-equivalents to hold it with?

Clearly, this required closer study. So the sponges were taken off the sea floor and removed to laboratory tanks, where things could be better controlled. The sponges were allowed to settle down for two days to recover their poise. Then the biologists began to cause them little alarms and heart-flutters, to see what would happen. They began to tap the side of the tank, or even go so far as to prod the sponges. Because sponges aren't as bright as all that and take a while to gather their wits, the sponges would wait for between twenty and fifty seconds and then catch their breaths suddenly. The reaction time for a sponge is certainly not up to that of, say a jet pilot!

Sponges Are People Too

Eventually, it occurred to the biologists to try giving the sponges electric shocks, to see if that would shake them up. Sure enough, the sponges got upset and 'stopped breathing' when shocked. Every time they were shocked, they held their breaths for between thirty and a hundred seconds.

By using electrical shocks like this it was possible to control the amount of stimulus so precisely that, by lowering the shock slowly, they eventually found a level which the sponge couldn't perceive any more because it was too faint. This gave 'a level of excitability of sponge tissue' which, because it could be expressed mathematically, gave the biologists untold delight.

But they still hadn't worked out how the sponges caught their breath. So they decided to measure the rate of transmission of a stimulus through a sponge, or its 'conduction velocity'. Ten sponges were used and the average rate was point twenty-two centimetres per second. That is slower than with nerves, but still pretty fast for one end of a sponge to learn that the other end has received an electric shock. In fact, it is too fast for chemical diffusion by passage of sea-water, and is considerably faster than previously observed 'contraction waves' in sponges, which traverse the body.

By now, the scientists were beginning to get a bit frantic about an explanation for all this. They cut away bits of sponge flap here and bits of sponge flap there to examine what was happening. They were looking for something like nerves or transmission lines. No luck.

They concluded that the sponges were emitting 'some kind of propagated signal or impulse which travels from the site of stimulation to widely dispersed effectors'. These 'effectors' then control the flow of water through the body wall. But the trouble is, no one knows either how the signal is propagated or what it is. Nor does anyone know how it is received, or how the sponge stops breathing – in fact, so bizarre are the conclusions that the scientists even suggest that for a sponge 'the entire conduction system could act as a single neuron'. What they are saying is that every sponge is a single giant nerve cell. That's a hell of a nerve! But it shows just how sensitive and emotional the little sponges are, and how one should be careful not to offend them. And one day, if their secrets are understood, it might mean a breakthrough in the concepts of science, for the biologists say they cannot rule out some quite novel, non-electrical signalling mechanism. And who knows, when discovered, it might have extraordinary and useful applications to human affairs. It could even result in a cure for shyness!

Sources: Dr I.D. Lawn; Bamfield Marine Station, Bamfield, B.C., Canada.

Dr G.O. Mackie and Dr G. Silver; Department of Biology, University of Victoria, B.C., Canada.

BLIND CHILDREN USE GEOMETRY INSTEAD OF EYES

Ingenious experiments with a blind child have shown that she was able to use the principles of geometry to a sophisticated degree to find her way along paths she had never previously taken, between objects in a room.

Kelli, the little girl used in the experiments, was two and a half years old. She had been born three months prematurely and shortly afterwards had become completely blind due to a disease. She took part in the experiments with her mother.

The experiment involved arranging four objects in a sort of diamond pattern around a room. One was the girl's mother sitting on a chair. The other three objects were a stack of pillows, a basket of toys and a table.

Kelli was brought into the room and taken to her mother. From there, she was taken to the stack of pillows and back to her mother, twice. Then she was taken to the basket of toys and back to her mother, twice. And finally, she was taken to the table and back to her mother, twice. Each time she was allowed to feel the object and become familiar with it.

Then Kelli was told, 'Go to the toy basket', or, 'Go to the pillows', and so on. She had never been to any of these things except on a direct route from where her mother was sitting. But she quite happily went rushing off from one to the other. She was filmed on videotape so that her movements could be closely studied. It was found that as she moved, she did not head 'ballistically', like a bullet straight for the target, but that she continually corrected and improved her angle as she went. Her success rate at finding her targets was extremely high, and two of her failures were only very narrow failures.

The psychologists decided to test to see whether she might have been doing all this high-powered navigating by using acoustic sound-cues from the shape of the room. So they took her out of the room and secretly moved all the objects round, preserving their angular relationship with each other exactly as before, but entirely changing their relationship with the room shape.

Allowed back into the room again after this secret change, and taken once more to her mother, Kelli was able just as easily to find her way from object to object. So, she was doing so because of the geometrical relationships between them, and not because of subtle hints from echoes of sound in the room itself.

What Kelli seems to have had in her head was a very deep subconscious understanding of Euclidean geometry, of relative distances and angles. She was actually using trigonometry to navigate! Knowing two sides and one angle of a triangle, she was able to 'know' what the other sides and angles were. This has considerable implications for the human brain, for learning, and for all of psychology. It also means that nobody has any right to fail a trigonometry exam at school! If a two and a half year old can do it without even thinking about it, there is no excuse.

Further experiments were carried on after this with blindfolded sighted people. Five three year old children who could see, and six adults who could see, were submitted to the same experiment as Kelli but with blindfolds over their eyes. The children had success rates similar to Kelli's, but the adults were a great deal better still. This indicates that geometry-on-the-brain is not just a characteristic of blind people. We all have the same abilities.

This certainly gives a new angle to things.

Source: Dr Barbara Landau, Dr Henry Gleitman, Dr Elizabeth Spelke; Department of Psychology, University of Pennsylvania, Philadelphia.

THE STILL INSIDE THE CELL

Alcoholics may not have known it, but there is a still inside every cell in the human body. The trouble is, it doesn't make whisky.

It is called the Golgi Apparatus. It is named after an Italian pathologist called Camillo Golgi, who discovered it in 1898 and later won the Nobel Prize.

For many decades scientists argued over whether the Golgi Apparatus existed or not. They thought that, because it stained so easily for mounting on slides and examination through a microscope, it was just an illusion somehow created by the staining. Then, as technology progressed, it was possible for the Golgi Apparatus to be seen three-dimensionally. And now the Golgi Apparatus has really come into its own. It is no longer possible to ignore its importance. Its purpose is akin to the process of distillation.

Cells contain quite a lot of what are called ER membranes, which make important proteins and enzymes which are essential to life. But a very tiny proportion of these substances, only one ten thousandth, in fact, are intended for 'export', that is, for use in other places. The Golgi Apparatus distils them and separates them out from all the other 9,999 substances. And one of the most important products of the 'Golgi Still' is deadly poison. (Not alcohol, but something worse!)

There are peculiar little things in cells called *lysosomes*, which contain the deadly poisons. Biochemists have described them as being the functional equivalent of the cyanide capsule that every good spy is supposed to carry. When things go wrong, the lysosome bursts and empties its contents into the cell, utterly destroying it. This could happen if a cell became injured or diseased, or when cells have outlived their usefulness. For instance, the tails of tadpoles are destroyed cell by cell by these little 'cyanide capsules' when a tadpole changes into a frog and wants to jettison its tail.

Apart from fatal poisons, material in the cell membrane is also produced by golgi distillation. The Golgi Apparatus when sliced looks like pieces of string laid out in a row. The materials go in at one end and out at the other. The first part filters and purifies by throwing out the unwanted 9,999 substances over and over again, until they are largely eliminated. The latter part gathers the special substances together, packages them safely in containers, and arranges the shipping to distant destinations.

The purification and refinement is so close to distillation that the professor who put the whole story together says quite clearly that, 'the purpose and principle of the Golgi may therefore be essentially that of fractional distillation, and the apparatus used may be fundamentally the same ... the front end is the distillation tower of this apparatus and ... the back end acts as the receiver, bleeding off the most refined product, often condensing it so that it can be more easily distributed in a concentrated form'. And these insights are the professor's distilled wisdom!

It's a pity, though, that some of the results of this cellular distillation are so deadly. Plainly it's a case, as the barmaid says, of 'choose your poison'.

Source: Professor James E. Rothman; Department of Biochemistry, Stanford University.

BEES KEEP PHOTOGRAPH ALBUMS

Bees take photographs, but not just to reminisce in their old age. For them, the photographs are daily working materials. In fact, a proper working bee will take more snaps in a day than the average photo journalist. And for the same reason: earning a living.

Two scientists had great success setting up experiments to catch the bees at it. The one way to make friends with a bee fast is to put out a bowl of sugar-water. This was duly done. In flew the bees, through an open window. The bowl was placed in an experimental room painted entirely white, with the window covered in white sheeting. Not too far from the bowl of sugar-water, a black cylinder was placed. This was intended as a 'landmark' for the bees.

Later, the bees came back hoping for some more sugar-water. But it had been taken away. The bees nevertheless hovered around the position in the same relation to the landmark.

As a result of this, the experimenters moved the landmark cylinder on subsequent occasions. Then the bees moved their searching location, keeping it in the same relationship to the landmark. This demonstrated that they had 'placed' the sugar-water in relation to the cylinder and were trying to reproduce or remember the scene as they had seen it previously.

The scientists then deviously changed the size of the landmark. It was then discovered that the bees flew in closer to the smaller cylinders, trying to get them to match the remembered size in their mind's eye. When a cylinder larger than the original was placed on the floor, the bees flew further away from it, obviously trying in the opposite fashion to match the original scene.

Then the scientists tried working with three cylinders, a set of three landmarks together. They gave the bees sugar-water placed in a certain position with the three cylinders, and then removed the sugar-water. The bees came back to where it had been. Then the scientists moved the three cylinders around the room, keeping their angular relationship constant. The bees always went to the point from which the three landmarks matched the original scene.

The scientists tried removing one of the three landmarks, and the bees did as well as they could using just the two remaining, which was pretty successful.

The scientists substituted black cubes of different sizes for the cylinders. But the bees didn't seem too perturbed. They went on concentrating on the angular relationships of the three. What they were looking for was evidently a scene as close as possible to what they remembered.

The conclusions drawn from all this by the two scientists suggested that bees 'store something like a two-dimensional snapshot of their surroundings taken from the food source. To return there, bees move so as to reduce discrepancies between the snapshot and their current retinal image ... (the bee) compares the image on its retina with the stored snapshot and moves until the two match'.

It would seem that bees keep photograph albums, in their heads, and that the sound of a bee camera is not a click but a buzz.

Source: Dr B.A. Cartwright, Dr T.S. Collett; School of Biological Sciences, University of Sussex.

EARTH BOOMS

Can the Earth speak? Yes! In fact, it has a distinctly booming voice.

Small earthquakes which are too small to be felt, can be *heard.* When there is a sudden upward thrusting of a piece of the Earth, it operates on the air to produce shock waves. This makes a boom. It is called 'the direct transmission of seismic energy from ground to air' and explains quite a lot of the booms which are heard, especially in California.

Mysterious booming sounds have been reported in many places over long periods of time. 'Earth booms' no doubt explain a lot of them, though there may be other explanations as well, of course, such as airplanes breaking the sound barrier high overhead and perhaps out of sight.

Minor earthquakes which could not be felt are definitely known to have produced earth booms in Landers, California, in the Mojave Desert, and around the Mammoth Lakes in California. In these cases, the booms were detected by scientists and thus confirmed.

In January and April 1980, minor earthquakes at Fontana and Berkeley, California, were not felt but were heard as booms. These earth booms were reported in the local newspapers at the time. In 1964, the great Alaska earthquake was heard as atmospheric booms and recorded on barographs many hundreds of miles away.

We are now learning to recognise the voice of the Earth, although we are still uncertain as to whether it is trying to tell us anything, or if it speaks in any 'Earth Language'.

In general, however, the safest maxim seems to be: when the earth speaks, run like hell!

Source: Dr Donald J. Stierman; Department of Earth Sciences, University of California, Riverside, California.

EATING GUNPOWDER

Some creatures have to eat the substances with which they defend themselves. It is rather like human beings eating gunpowder and then stuffing rocks in their mouths and shooting their enemies with explosive belches.

Things are almost as drastic with a Florida moth which eats little cochineal insects. The cochineals live on cacti and are called *Coccus cacti*, not surprisingly. The moths which eat the *cocci* are called *coccidivora*, or 'cocci devourers'.

The cochineal is quite a famous little insect. For untold time in Mexico it produced the famous cochineal dye. Aztecs and Toltecs, just like their modern successors, would walk through specially planted cactus gardens and brush the cochineals into straw baskets, leaving enough behind to breed for the next year's crop. This would be done just before the rainy season. Then, the gathered insects would either be killed by being thrown into boiling water, or placed in linen bags and baked in ovens. The latter method preserved the peculiar white down with which many cochineals are covered (it is thought to be a protection against predators, but it doesn't protect them against man).

Carminic acid is the active dye in the cochineal. Artists will be familiar with the term carmine, which is a paint pigment derived from cochineal. This substance produces a bewildering variety of different colours depending on which mineral 'fixer' or 'mordant' is used with it. On wool, cochineal can give a vivid crimson colour, or a very fiery scarlet. Or it can yield a fine purple, or a lilac, or even a slate colour.

Carminic acid is a very potent insect repellent, especially repulsive to ants, who won't touch it and run a mile. The cochineal insects are stuffed full of this acid, and very few creatures would be ill-advised enough to want to try to eat them. But the 'coccus devourer' moth, in its larva stage, devours them greedily. It takes all the cochineal material it eats and shunts it aside into a special fore-gut compartment, which is tightly constricted and sealed off by muscles.

Ants would be natural predators and enemies of the larvae of the 'coccus devourers' in the normal run of things. However, the 'coccus devourers' aren't so called for nothing. When ants see the larvae and go rushing up to them to take a bite, the larvae release cochineal matter from their fore-gut compartments and vomit it all over the ants, who turn and run like crazy, limping and hopping along like wounded soldiers, and just as covered in red sticky stuff as any soldiers ever were drenched in blood from a battlefield.

Studies of the vomited cochineal matter have revealed that it has higher concentrations of carminic acid in it than the cochineal insects have themselves! So the 'coccus devourers' have been eating cochineal insects to good purpose, extracting the poisonous material, and concentrating it further. The larvae, when threatened, even smear their bodies all over with the cochineal, just to be really revolting, and no insect will take the slightest nibble of any areas so besmeared.

Not only have the moth larvae managed to crash through the defensive chemical barrier of the cochineals, as the scientists put it who discovered it all, but they have robbed the cochineals of their weapon and even improved it. Clearly, with the moth larvae, it's a case of 'fighting makes me sick!'

Sources: Dr Thomas Eisner, Dr Stephen Nowicki; Section of Neurobiology and Behaviour, Cornell University.

Dr Michael Goetz, Dr Jerrold Meinwald; Department of Chemistry, Cornell University.

THE LOCUST PLAGUES OF THE UNITED STATES

Locust plagues happen at such long intervals that people tend to forget them. Children, who aren't old enough to remember one, are struck dumb with horror and turn to their Bibles to read about the Old Testament prophets. What doom is portended?

Having found myself in the middle of a particularly severe locust plague once, I can assure the reader that it is an indescribable experience. The locusts are ankle-deep on all the streets, no matter how fast you sweep them up with brooms. It is impossible to walk without the most hideous and sickening crunching sounds at every step. The creatures are all over you – they fly into your face, sit in your hair, grip at your elbows and even crash into each other in the air in front of you, making a thud which is unlike any other thud you have ever heard, and far more disturbing. They seem to claw at you with their little feet and spindly legs. Surely, it was locusts that inspired ancient peoples to imagine 'demons'!

Because the locust plagues are so far apart, scientists have difficulty studying them. By the time a scientist has waited perhaps seventeen years for a recurrence, he may have retired. It has, therefore, been a slow business getting the information together to understand what is happening.

In the past twenty years some progress has been made. Two scientists named Dybas and Davis seem to have studied the very same locust plague I experienced – the plague of the summer of 1962 in New Jersey. In a report published that year, they informed their, no doubt horrified, scientific readers that the densities of living matter present in the form of locusts in a plague exceed the densities of living matter of any other type of creature anywhere on the Earth. Anyone who has seen it can believe it. They counted the locusts and found that three hundred of them were emerging from every square yard of ground. That means that there were one and a half million locusts *per acre*. No wonder everyone in New Jersey in the summer of 1962 was in a jangled state of nerves: we were in the middle of the equivalent of a biological hydrogen bomb explosion. It was the most intense and powerful outburst of living flesh from the ground that can happen on our planet. Fortunately, it only lasted for about three weeks, and then came the bonfires, the description of the smell of which I shall spare the reader.

Recently, the first study has appeared which quantifies the amount of damage these locusts do to the trees. The locusts are supposed to have either a seventeen year cycle or a thirteen year cycle, though comparing one state with another, they don't seem to all 'come out' in the same year. The locusts come out of the ground, mate, and the females lay their eggs in the twigs of non-coniferous trees. When the eggs hatch (the adults having died and their carcasses left to rot) little nymphs emerge and fall to the ground. They dig down into the ground and start eating the roots of the trees. At first they eat tiny rootlets which are bite-sized. When they finish, having killed the rootlet entirely and got as much out of it as they can, they move on to the next rootlet. They gradually move further and further down into the ground over the years, eventually reaching a depth of about two feet. By that time, they no longer inhabit little tunnels, but have made oval-shaped cells for themselves, tucked up against a large root on which they then feed. These nymphs die from three main causes: moles eat them (so never kill a mole, he is the tree's best friend); they can die from over-competition by neighbouring nymphs for food, or they can simply starve if they get unrewarding roots.

It is the baby nymphs who do the most damage. Large roots which the older nymphs eat can withstand being gnawed at. But the killing of myriad tiny rootlets by the infant nymphs does enormous damage to the trees. Some detailed studies have been done and it has now been demonstrated that trees can withstand one or two nymphs nibbling at them without showing much damage. But from three upwards, things begin to change. Tree growth-rings have been studied for trees infested and neighbouring trees non-infested. It was found that feeding locust nymphs are capable of reducing tree growth by as much as thirty per cent. This means that these locusts are responsible for diminishing American wood growth to a gigantic extent. It is an understatement, therefore, to say that they are a serious pest. Plague is the only word for them. And one wonders what Moses would think of it all.

Source: Dr Richard Karban; Department of Biology, University of Pennsylvania, Philadelphia.

IS THERE A RING AROUND THE SUN?

Everyone knows about the rings around the planet Saturn. Space probes have revealed similar, though less dramatic, rings around the planets Jupiter and Uranus. So several astronomers, carried away by it all, have suggested that the Sun itself has such a ring around it, and if we look hard enough during a full solar eclipse, we might be able to see it.

Rings around heavenly bodies will no doubt be fully understood one day, but at the moment we are still learning about them. The latest discoveries about the Saturn ring system have revealed that it consists of a huge stretch of rings covering the equivalent distance as from the Earth to the Moon, but they are only thirty feet thick! If the Saturn rings were shrunk to the size of your hand, they would be so thin as to be invisible. It is incredible to think that something so extensive has absolutely minimal thickness.

What leads the astronomers to predict a ring around the Sun? After all, there don't seem to be rings around the Earth, Mars, Venus, or Mercury. So, why the Sun? There is a kind of ring around the Sun already, so obvious it almost doesn't occur to us: the planets themselves, largely lying in a plane, are a kind of extended ring. But that isn't what the astronomers are referring to.

Basically, the idea is that as the planets show a certain regularity in their spacing outwards from the Sun, according to a pattern sometimes expressed as 'Bode's Law', there should be – according to a later version of it known as 'Blagg's Law' – a 'planet' of some kind inside the orbit of Mercury, the nearest planet to the Sun. This planet should be just inside the limit at which gravitational tidal forces would cause such a body to break up. So it is computed that if it existed, it would exist as a cluster of rubble, made of rock, which would appear as a ring around the Sun. It would be somewhere about four solar radii distant from the Sun.

The astronomers compute that the pieces of rock would be about ten kilometres across and that there would be at most a million of them. They say that such objects would be observable, in practice, only during a total eclipse of the Sun.

The above suggestion was published in November 1979. Three months later, a group of seven astronomers at Bangalore in India attempted to find the ring around the Sun during a total eclipse. A year later they published their findings. They were unable to see the proposed ring and set limits on its characteristics. But they didn't rule it out, and what they did obtain from their observations and computations, 'makes the problem more enigmatic and interesting for future measurements'. Sounds like it isn't a dead-ringer, but it isn't quite a dead ring, either.

Sources: Dr K. and A. Brecher; Department of Astronomy, Boston University.

Dr P. Morrison; Department of Physics, M.I.T.

Dr I. Wasserman; Centre for Radiophysics and Space Research, Cornell University.

Drs U.R. Rao, T.K. Alex, V.S. Iyengar, K. Kasturirangan, T.M.K. Marar, R.S. Mathur, and D.P. Sharma; ISRO Satellite Centre, Bangalore, India.

MAIDEN AUNTS AMONG FOXES

Vixens don't all have cubs. But the 'unmarried' ones can still form part of the family and take care of the offspring. In fact, fox maiden aunts are an integral part of fox society.

Wild foxes were studied in Oxfordshire, and sixty per cent of the vixens were discovered to be barren. Only one, or at the most, two, vixens had cubs in every fox family. Foxes seem to live in groups of four or five, a fox family consists essentially of a male and his mate, and some 'maiden aunts'. They are of course joined by the cubs for some months who stay on if they are females, but males go off on their own in their first year.

Radio tracking foxes in the wild got to be rather exhausting after a while, so the zoologist concerned decided to take some fox families home with him. He captured and enclosed two separate fox families and gave them a thousand square metres each. Every autumn he removed the male cubs, just as would happen in Nature. He also removed females at random ages, to avoid over-crowding. He studied these captive fox families for a total of eight years.

During the eight years, the female foxes present could have had as many as twenty-seven litters if they had all been mothers. In fact, only seven litters were born. This meant that there was only a thirty-two per cent reproductive rate amongst the foxes. And every year, only one vixen per family gave birth (except for a single year, to be discussed later). The rest of the female foxes not only did not give birth, but during the mating season were entirely ignored by the male fox. What is more, the maiden aunts were lower down in the social hierarchy than the mother fox. The mother was very obviously dominant, bossing all the other females around and accepting their kowtows. These maiden aunts were themselves ranged in a descending hierarchy of status. The lowest of them all acted as a kind of scapegoat when the mother was in a bad temper. For the first three weeks after the birth of her cubs, the mother would furiously attack the lowest-ranking female and get rid of her frustrations. (She didn't attack the middle-ranking ones.)

The maiden aunts were extremely helpful and fetched food for the cubs, though never for the mother herself. They groomed the cubs and looked after them as if they were their own flesh and blood. And, of course, in a sense they were, since they were all in effect nephews or nieces. This relationship is thought to be an important factor in binding the maiden aunts to the cubs – they seem to sense that they are related in some way.

If wild foxes came sniffing up to the fences round the captive fox family enclosures, the maiden aunts went running out and attacked them and chased them away in those instances when the mother was busy, such as tending her cubs or heavily pregnant. And a revealing incident occurred one year in one of the captive families. The mother suffered a serious injury to a front paw when her cubs were only four weeks old. For fourteen of the next sixteen days, she was entirely unable to tend her cubs. Her duties were taken over by the all-too willing maiden aunts, and the cubs were perfectly all right.

There was one year in which a fox family produced two mothers. On this exceptional occasion, one vixen was declining in status and the other was ascending, so the two were in the process of exchanging dominant roles. Among foxes it is the dominant female who seems to have the exclusive privilege of mating. But at this point, the dominance was temporarily unclear, so both were mated. Let us call these two vixens Beverley and Wanda.

Wanda had been the dominant vixen for some time, but began to sink and lose her nerve. So Beverley made a grab for power, and during the time this change-over was happening, both became sexually attractive to the male fox, and the unusual double-mating occurred. Beverley had her cubs thirteen days earlier than Wanda. Although Wanda had been a successful mother the previous year, when she finally had her cubs in this particular year, she seemed obsessively guilt-ridden. She repeatedly picked up her cubs and carried them around the enclosure, uttering submissive shrieks and repeatedly taking a submissive posture. 'Forgive me, forgive me!' she seemed to be pleading to the now-dominant Beverley. Sometimes she would do this spontaneously, but if Beverley approached her, it triggered her off.

After eight days, Wanda had let her cubs die. She took the corpse of the last cub and carried it to the entrance of Beverley's earth and made submissive postures, begging for her forgiveness and esteem. Wanda then visited Beverley's cubs no fewer than ten times during ten minutes, and Beverley made what appears to have been a gesture of magnanimity by visiting Wanda's empty earth three times (possibly to check that all the rival cubs were really dead). After this, Beverley allowed Wanda to nurse her cubs and Wanda settled down quite happily, with the two of them acting as joint mothers of Beverley's offspring.

Source: Dr D.W. Macdonald; Department of Zoology, Oxford University.

THE CHAP WHO REALLY GETS AROUND

Although he 'circulates' a bit too literally, there is a lowly creature of the intertidal sands who really does get around. He has no English name, but we can christen him the 'somersaulting squill'. He is one of the *stomatopods*, which are an order of the crustaceans better known as squills. Crustaceans are not unfamiliar to most people, since they include the crabs, lobsters, and shrimps.

The somersaulting squill lives up and down the coasts of Southern and Central America, and was studied in the sandy beaches of the Canal Zone by a zoologist from California. The little fellow lives in the sand. He is tiny, only twenty-three millimetres long, in fact. He has a long, flat body, looking a bit like a squashed centipede with fewer legs. He burrows holes for himself in the sand, head first, then does a reverse turn and crawls up with his mouth full of sand, which he dumps and then goes back to dig some more. After he has dug his cosy burrow, he sits in it underwater with just his eyes and antennae protruding, waiting for prey to come along.

The little fellow likes swimming, but when he is stranded on the wet sand when the tide goes out, his legs aren't strong enough to lift him off the sand, and he can't walk properly. Fortunately, he doesn't have the rounded back that most crustaceans have, so what he does is turn himself over on his back, lift his tail into the air and throw it over his head, form a loop, and then flip himself over in a somersault.

When they really get going, the somersaulting squill can do as many as forty successive somersaults at great speed, and usually this means they reach the water before they are completely exhausted. Because the somersaults are backward ones, this means that as they flip their heads over, they can point themselves in a new direction. On a downward slope they tend to descend like a rolling hoop, pulled by gravity.

A somersaulting squill can climb a slope of ten degrees on a beach, and in a laboratory (because I hardly need tell you the zoologist took some home with him) they can climb an incline of thirty degrees. This is the only known creature on Earth which does somersaults uphill, although there is a spider which forms its body and legs into a sphere and rolls down sand dunes. But the somersaulting squill's actions are quite unique – falling head over heeels doesn't seem to improve its love life. They seem to live alone, and have no one to look after them, clean the burrow, or applaud when they do particularly brilliant somersaults.

Source: Dr Roy L. Caldwell; Department of Zoology, University of California at Berkeley.

WELL, I'LL BE A MONKEY'S BROTHER

Being a monkey's brother can be a real advantage. It can be even better than being a monkey's uncle. It gives you status, lots of girls, perks, and a high time. You always have a friend to fall back on, and with any luck you can even become Top Banana.

Who is it who has such an easy ride through life? The rhesus monkey. It's nepotism all the way with them. Male rhesus monkeys always leave the groups into which they are born. The groups are quite large, between one and two hundred monkeys, and they often come in contact with each other. It's therefore no problem to switch groups if you feel like it, and for some reason all the rhesus young males feel like it. But what group to join? It's like the problem of too many societies on campus, or finding yourself in one of those streets with ten restaurants door to door and having to choose one when the menus are all the same. You like steak? They all serve it the same. So you end up choosing the one with the prettiest waitress.

Rhesus monkeys end up choosing the group which has one of their brothers in it. Several groups of them were studied on a small 100-acre island off the coast of Puerto Rico. The monkeys were all taken there and let loose, after having had letters and numbers tattooed on them and having been entered into genealogy charts showing who was related to whom. These studies took place over a period of nearly twenty years.

It was discovered that monkeys welcomed their brothers into their groups with them and that brothers spent far more time together than with any other males. They treated each other with respect and affection, even going so far as to refrain from grabbing hold of each other and tugging when they were copulating with their girl friends. (Monkey manners are so appalling that this is a real plus, apparently.)

The usual pattern for a male wanting to join a new group is to go for the group with the females. The more females there are, the more eager the males are to join up and get some of the action. But brothers held more appeal even than females. When there were brothers available, the rhesus monkeys passed up the tempting opportunity for more sex, and joined up with the brothers in groups with fewer females. But they were soon compensated for their filial devotion. They went shooting up the hierarchies of importance and showed all the advantages of well-connected young men-about-the-trees. So they were apparently either motivated by irresistible family feelings, or saw long-term advantage. But when big brother met with a sticky end, little brother immediately suffered the consequences. One such fellow, whose two older brothers died, fell immediately four levels in the social hierarchy because he had lost his protectors. This was the most drastic social fall from grace seen in the twenty years, and showed how important it is to be part of a family. Isn't it touching?

Source: Dr D.B. Meikle and Dr S.H. Vessey; Miami University, Oxford, Ohio, and Bowling Green State University, Ohio.

THE FORESTS THAT ARE LIKE THE SEA

The sea is always moving, ever changing, and yet it remains the same. Forests too can be like this. Studies of fir forests by a forestry expert have led him to describe his forests in a manner which reads like the pronouncements of Chinese philosophy: 'At any given point the wave-regenerated forest is in a state of constant change as it degenerates, regenerates, matures, and degenerates again in the endless cycle ... all stages of degeneration and regeneration are present in the forest at all times ... the forest as a whole remains relatively constant ... in what might be termed a steady-state'.

Fir forests have actual waves that sweep through them at a more or less constant rate, and at any given time about the same number of trees are dying as are growing. The waves have actually been photographed, and are quite striking and obvious, once one is told what they are. Seen from a distance, they appear as curving bands of dead trees. At a casual glance they may look like rows that have been blasted by a storm. But upon closer inspection, it is found that they are always in the same pattern: mature forest suddenly ending in dead trees, followed by young saplings and taller and taller saplings which extend up to a mature height. Looked at from the side, it is remarkably like a sea wave.

These 'regeneration waves' as they have been named, have been shown to sweep through the forests as if the trees were liquid. The reason they weren't noticed before is that they take so long, and move so slowly in terms of a human lifetime. A typical forest wave in a fir forest will sweep along at the rate of about three metres per year, which is not noticeable to someone standing there! The waves are regular, and every sixty years a wave will sweep across the same spot in the forest. Since the fir trees (the study concentrated on balsam firs) live for about sixty to eighty years, the wave speeds are in synchronisation with the tree lifespans.

Sometimes the wave edges are so obvious that everyone remarks upon them. This is true in Japan, where in the dark green forest on the slope of Mount Shimagare several whitish horizontal stripes running in parallel with each other can be seen from a distance so distinctly that the mountain has been named (in Japanese) 'mountain with dead tree strips'. In America there are some incredibly striking wave patterns to be seen in fir forests in New York, New Hampshire and Maine. Some of the best appear on Mount Katahdin in Maine.

The forest is thus a vast entity, through which cyclic waves of death, regeneration and maturation constantly move. It is always in flux, and yet always the same. It gives one pause to think that trees growing collectively out of the soil can behave so like the waters of the deep. A forest is like a sea which has frozen solid, its waves slowed down from seconds to the creeping progress of years. But it is as if only the time-scale has changed, and if you were to cause the forest-waves to speed up by several thousandfold, the forest might liquify and slosh against the mountain peaks, sending sprays of pine essence into the sky.

Source: Dr Douglas G. Sprugel; Department of Forestry, Michigan State University, East Lansing, Michigan.

Dr F.H. Bormann; School of Forestry and Environmental Studies, Yale University.

ARE THERE TWO SUNS IN OUR SOLAR SYSTEM?

This may sound like a crazy question – if there were two suns in our solar system, surely we could see the other one, couldn't we? Wouldn't we only have to look up at the sky and say 'There it is'?

The answer is no. A second sun in our solar system has been proposed by a respected astronomer who says that it would be far too faint to be seen by the naked eye, even at night. He has in mind a very tiny little star which he believes he may have detected by its effects on some other objects in the sky. He proposes that we try to find it, or otherwise that we try to find some means of explaining his sighting.

The suggestion was first made public in November 1977, by Dr E.R. Harrison. For some years previously he had been studying the characteristics of pulsars, which are highly compact pulsing stars in our galaxy. Pulsars throughout the sky tend to slow down very gradually in their pulsing. But Harrison noticed that six pulsars, which were not slowing down as much as all the others, were grouped together in the same tiny region of the sky. He thought this very odd and decided that some special influence must be working away at them to cause this.

Harrison worked out mathematically that the effect on the pulsars could be explained if the centre of gravity of our solar system was being shifted towards them, in the direction of the galactic centre. The only way he could imagine this happening would be if there were another star in our solar system which was found in that direction.

This sparked off a lively debate amongst astronomers all over the world. Harrison had suggested that the star must be a dwarf, but he didn't know what kind of dwarf – white, red, neutron, black hole, or what? Neutron stars and black holes are massive and dense objects, confined to tiny size. Various astronomers published articles about Harrison's theory, and it emerged that if there really is a star such as he suggests near the Sun, it must be one of the highly compressed ones such as the neutron star or the black hole. An astronomer called Wright suggested a black hole at a distance of one light year, and thought that it could be detected if a search were made for radiation of a certain kind.

Two Dutch astronomers did not like these ideas at all, and wrote an article insisting that the strange characteristics of the six pulsars must have another explanation. They said they had computed that if such a star were there, it could not have less than one third of the mass of the Sun, but that would make it too bright. What they really ruled out by their calculations was a normal, non-compacted star. A French astronomer then produced an article insisting that such a star, if it were a neutron star or a black hole, 'could have remained undetected until now'. He thought it must only be a temporary companion to our Sun, which was moving past us through space. He thought it possible that such an encounter might currently be in progress, and that it was strictly a high-speed, transient affair. In December 1979, in apparently the last published comment on this matter in a scientific journal, an astronomer in Austria expressed scepticism, saying that the expected effects on comets were not commonly enough observed to justify the suggestion of a second sun.

Since then, nothing further seems to have happened one way or another. As far as Harrison is concerned, he apparently still believes the second sun may very well be in our solar system, as do the people who supported him. Its existence has not yet been disproved, but if there is a black hole zipping by our Sun, let's hope it misses us!

If one day we look up and see two suns in the sky, we won't be seeing double. We'll be seeing 'Harrison's Sun' which has moved a little closer. If it exists ...

Sources: Dr E.R. Harrison; Department of Physics and Astronomy, University of Massachusetts, Amherst.
Dr E.L. Wright; Department of Physics, M.I.T.

Dr H.F. Heinrichs and Dr R.F.A. Staller; Astronomical Institute, University of Amsterdam.

Dr Serge Pineault; Department of Geophysics and Astronomy, University of British Columbia, Vancouver.

Dr Daniel Wilkins; Institute of Theoretical Physics, University of Vienna.

HOW DO WE KNOW WHERE WE ARE GOING?

You may think, as you hurtle along the road in your fast car, that you know where you are going. But are you absolutely sure? What about when you are piloting an airplane? Running? Or, for that matter, crawling along the ground? How do you really know the direction in which you are moving?

If you were driving, you might say that the trees which were closer moved faster than the trees further away. The trees, therefore, at the side of your visual field zipped past, while the ones further down the road were slowly building up speed, but by the time they reached you, they too would have accelerated and zipped past you. The same is true for cars coming along the opposite lane. A car in the distance moves slowly, then when it passes you, you realise it is doing seventy mph.

The way we judge the distance we are moving has been assumed to be due to the different speeds at which objects at varying distances are apparently moving past us. This is known as the 'expanding pattern of visual flow' in which 'image velocities' of things we see vary according to distance. If we are moving towards a fence, the posts nearest the point we are aiming at are supposed to move towards us faster than the posts further off to each side. It seems common sense.

But no! This idea has now been overthrown by two Canadian physiologists. They won't let us get away with this any more. They had their doubts, and they have now proved them.

It seems that if you move the direction of your gaze to the side slightly, and do not peer exactly at your destination when you are moving, the apparent speeds of the objects in your visual field will alter in such a way that, according to the notion discussed a moment ago, you would appear to be moving in the direction you were looking, rather than in the direction you were really moving. The scientists found a devious way of proving their point.

They took photos of a girl standing against a background of a wall covered in polka dots. Each photographic plate had multiple exposures, as the girl stood very still and the camera moved in towards the wall. The first multiple-exposure photo was taken with the camera pointing and moving directly toward the girl's head. Not only was the girl's head in focus, but only the polka dots immediately surrounding her head, like a halo, were dot-shaped. The surrounding dots on the rest of the wall were stretched and elongated like arrows of increasing length as one looked outwards from the girl's head. This is the sort of thing everyone would expect.

But the second multiple-exposure photo was the interesting one. In this one the camera was moving in exactly the same way toward the girl's head, but it was pointed slightly to one side, so that it was gazing at the wall beside the head. Its direction of motion was identical, but its gaze was shifted.

And this was the surprise. In the resulting photo, the girl was quite blurred and the polka dots to one side of her were in focus, getting progressively less in focus as one looked towards the girl and away from the centre of gaze.

This provided dramatic evidence that knowing where you are going is not as simple as everyone thought. The expanding, bursting pattern of visual flow as you move has its centre at the centre of your gaze, not at the point of your destination.

How, then, do we work out where we are going? Athletes, for instance, often have to know precisely. In fact, an athlete in a fast-running ball game like basketball might have been able to save the scientists the trouble of complicated experiments. He could have told them that when you rush down the court dribbling and looking at the far basket, all moving objects, such as other players, appear to diverge outwards from that distant goal, rather than from the frequently shifting destination of your changing path. But it is now scientifically established. So where do we go from here?

The scientists, of course, came up with another test. They got a group of subjects to agree to an experiment. They prepared a programme with a computer and a visual simulator television screen. It simulated the view a person would have travelling in an automobile at fifty-five kilometres per hour towards a wall seventy-six metres away.

By presenting various different centres of expanding flow, along with different positions of a dark bar and different centres of gaze (left or right of the dark bar), the triumphant answer came back: we judge where we are going from that point on the retina of the eye at which maximum rate of change of magnification occurs. This has nothing to do with the expanding flow pattern, and is independent of where we rest our gaze. As long as the point of destination is in the visual field somewhere, the rate of change of magnification of that part of the visual field will be greatest, whatever we are looking at. It therefore seems that some part of the brain is not distracted by what we are looking at, and concentrates intently on a mathematical quantity: rate of change of the magnification within the image field. Is there somebody sitting up there with a pencil and paper and a sliderule?

Those of us who didn't like mathematics at school and never dreamt of becoming engineers may well be surprised to learn that we have a 'slave' in the brain who computes this kind of thing without bothering our conscious minds in the least. Let's hope he never goes on strike or demands higher wages.

Source: Dr D. Regan, Dr K.I. Beverley; Department of Physiology and Biophysics, Dalhousie University, Halifax, Nova Scotia.

WHY OUR SKINS ARE BLACK AND WHITE

Why are black people black? Until recently there was actually no sensible reason why they should be, according to any scientific theory. Black skin is white skin with lots of extra pigmentation in it. Why is the pigmentation there?

Now at last an explanation may have been found. It has to do with essential vitamins in the human body and the need to protect them from destruction by the sun's rays. There are several vitamins necessary to life and health which are badly affected by sunlight, particularly by ultra-violet sunlight. Experiments have been done with these substances which prove that human beings exposed to too much solar radiation would not have enough either to remain healthy or, in some cases, to remain alive at all.

Riboflavin, better known as Vitamin B-2, carries oxygen to the cells. A deficiency of it will result in the disease known as pellagra. A less serious shortage of it will bring dermatitis, sores at the corners of the mouth and nostrils, trouble with the cornea of the eye, and other problems. The vitamin is obtained from green leaves, milk and eggs. Without its help in producing energy in the body by the burning of our food, we are in real trouble. It is a strange substance to look at: on its own it comes in thin crystals of orange-yellow needles, but when dissolved in water it is fluorescent and actually glows a strong yellow-green. However, upon exposure to sunlight, this vitamin is very quickly destroyed.

Vitamin E comes from green leaves, but especially from wheat germ oil, and is a solid alcohol whose exact requirements by human beings is unknown. But it is certainly essential, because experimental animals deprived of it have developed muscular weakness and female rats become unable to bear live babies. This vitamin is also absolutely essential for successful reproduction; it is sometimes for this reason called the anti-sterility vitamin. Vitamin E is also vulnerable to destruction by sunlight.

But the crucial vitamin which was studied in the experiments was folic acid. A severe deficiency of it can be fatal. If babies and children don't have enough of it, their growth is retarded, their blood will be abnormal, and they will be sickly. But the most common deficiency of folic acid is in pregnant women. It is estimated that anywhere between ten and sixty per cent of all pregnant women in the world do not have enough folic acid in their bodies. In extreme situations this can cause the pregnant woman to die. A large number of miscarriages, stillborn babies, congenital abnormalities in children, and other problems connected with pregnancy occur because of deficiency in folic acid. Since such a vast number of pregnant woman suffer from deficiency of this vitamin, the problem is one of immense concern to large numbers of people. In the West, particularly in health-conscious America (where health is taken so much more earnestly than in most of Europe), pregnant women make certain that they take folic acid vitamin pills during pregnancy. There can be no doubt that this has prevented many serious pregnancy complications. It is only to be regretted that governmental health authorities all over the world do not warn pregnant women of the folic acid problem and that progress is so slow in the spreading of awareness of it.

Concentrating on the folic acid in their experiments, two scientists decided to test their theories about skin colour in black people. They took normal blood plasma and exposed it to ultra-violet light rays. After only one hour (and they tested that no evaporation had occurred), the folic acid in the blood plasma had dropped by between thirty and fifty per cent! Whereas the folic acid in some plasma in a control experiment, which had not been exposed to the light, had not declined at all. This proved conclusively that folic acid in the form in which it exists in the human bloodstream and tissues is drastically susceptible to destruction by light.

It was necessary to carry the experiments further in order to be certain that the human body did not have some compensating action so that the folic acid would not actually disintegrate inside the body as it did in the jars. The experimenters found ten patients undergoing therapy for skin disorders which involved being exposed to ultra-violet light for half an hour once or twice a week for at least three months. (Ultra-violet lamps are known to help dermatological problems.)

The scientists tested the blood of the skin patients and compared the results with the blood of sixty-four normal healthy white-skinned people. There were significant differences in folic acid content. With five skin patients in particular, whose skin ailments had largely been cured by the ultra-violet light, the low levels of folic acid in their blood showed that it was not connected with skin problems alone because the skin problem had largely gone. Not only does this show that all people taking ultra-violet light treatment should routinely be given vitamin pills containing folic acid, but it proved that ultra-violet light – which, of course, one gets also from sunlight – not only destroys folic acid in jars, but probably very quickly in the human body by shining on the skin.

It appears, therefore, that black people had to develop the concentration of dark pigment in their skins in order to screen out the intense tropical rays of the sun, so that they could survive. If not, the folic acid and the other vitamins mentioned previously would have been destroyed so often that most of them would have died and the human race may not have been able to continue to inhabit places like Africa. And, since it is thought that the earliest humans came from Africa those people must either have been covered in hair, or black-skinned.

Black skins, then, are necessary for people living in intense sunlight to save their lives! People living in the cold regions of the Earth, especially places like Iceland and Sweden, need not go to the trouble of having pigment in their skins. The danger to them of sunlight destroying their vitamins is so remote that they can develop skins entirely free from pigment.

Using the above criteria, the evolution of various peoples can be looked at afresh. It could be argued, to take an example, that the supposed Mediterranean origins of European Ashkenazy Jews seems rather doubtful, since these people are usually rather pale and betray little evidence of ancestors from sun-baked regions of the Middle East. They may well have different origins such as those suggested by the author Arthur Koestler in his book *The Thirteenth Tribe.* Otherwise, how could they have lost their skin pigmentation in only a few centuries? If we compare them to the Eskimos, who have retained unnecessary skin pigmentation for probably more centuries in the frozen North, the argument for European Jews having Middle Eastern origins becomes rather weak.

Living in the tropics is exceedingly dangerous from the point of view of vitamin destruction in the body and deficiency of folic acid is a serious problem there in any case. It is often diagnosed as anaemia. Folic acid is also needed more in the body when combating malaria, so that even more severe anaemia may be caused in malaria victims.

The World Health Organization's recommended daily intake of folic acid in the diet (which is probably a conservative estimate) is ten times more than the amount of folic acid in the typical tropical diet as determined by various nutritional surveys. Thus, not only are the vitamins being destroyed by the sunlight, and sucked dry by malaria, they are being eaten about ten times less than required in the diets of the people most endangered.

This is a major world health problem.

Source: Dr Richard F. Branda, Dr John W. Eaton; Department of Medicine, University of Minnesota, Minneapolis, Minnesota.

CAN SEX KILL?

Experiments have shown that sexual activity shortens the lifespans of male fruitflies. However, the experiments must have been pretty exciting for the group indulging in all the sexual feats, as sexual activity was manipulated by supplying individual males with receptive virgin females at a rate of either one or eight virgins a day!

A control group for comparison was made up of male fruitflies who merely sat around with no females to entertain them. Another comparison group of males were presented with females who had already mated - such females don't mate again for some time.

Both the control groups lived for the same length of time; having no females around and having non-available females around amounted to the same thing. Just looking at them didn't have any effect on longevity.

The normal lifespan for a fruitfly is about sixty-five days. The male fruitflies who were given one virgin per day only lived on average fifty-six days. But the ones who were given eight virgins per day lived a mere forty days! So, clearly, the more sex they engaged in, the more quickly the creatures were wearing themselves out - performing themselves to death.

Apparently the fruitflies were spending a lot of energy on the production of sperm and seminal fluid and muscular action associated with mating itself. Of course, we do all that too. And it will not be lost on the reader that women live longer than men in the human species. The question which might arise, therefore, is: could we males possibly be killing ourselves with sex?

Can Sex Kill?

In order to try and figure this out I made a few calculations. They are offered only in fun, have nothing to do with the scientists' work on the fruitflies and are not intended to be reliable. But let us assume that the average man makes love at least three times a week, for the forty-five years between the ages of twenty and sixty-five. That is 156 times a year, a total of 7,020 times in his lifetime. This is equivalent to making love every day for nineteen years. Now, we all know that men live about five years less than women. I don't have the precise figures, but this is only in fun and there is no need to be precise. So let us accept the figure of five years.

If the average man has engaged in the equivalent of nineteen years of daily sexual ejaculations and he lives five years less than women for no other known reason, is it possible that every sexual ejaculation costs a fraction of the man's lifespan through the use of energy, and stress and strain on the organism? If the above rough figures were valid, they would mean that every ejaculation might take .263 of a day off a man's life!

Source (for fruitflies only!): Dr Linda Partridge, Dr Marion Farquhar; Department of Zoology, University of Edinburgh.

SPIDERS THAT SEND TELEGRAMS

There are some nocturnal wandering spiders in the tropics which do not live in webs. So, when they want to find each other, how do they know where to look? If there is no web, then they are never either at home or away from home. But this is clearly not satisfactory for spider social life. Or, to put it more urgently, for spider sex life.

Experiments have been done with some of these spiders, called *Cupiennius salei*, who customarily live either on banana plants or agave plants. I don't know whether you have ever tried living on a banana plant. Perhaps you have. But for those readers who have no experience of it, I should explain that it can be very difficult to find your girlfriend on such a plant because she is almost bound to be out of sight. Banana plants have a central stem from which a group of long drooping leaves hangs down on separate stalks, and some are quite large, so you could easily spend all day searching for her.

If you are a spider, what do you do? Do you cry out for her? If you did, your voice might be lost in the rumble of the jungle. Do you therefore crawl down all the leaves in turn, returning to the central stem each time to cross over and pass down the next leaf? This is exhausting, and does nothing to stoke the fires of passion.

What is a spider to do in a situation like this? The species must go on. Reproduction must occur. The female must be found.

Two German arachnologists (wonderful word!) experimenting with spiders on the banana plants in a laboratory, drowned out the spiders' calls and shrieks by playing random noise out of loud speakers up to eighty-six decibels. The male and female spiders were thus absolutely unable to call to one another. By this means the scientists confirmed that 'airborne sound is not needed for courtship communication in this species', since the spiders found each other anyway, despite the loud speakers.

So how did the spiders manage it? Did they call on Western Union and send each other telegrams with directions on how to find one another? Yes! That is more or less what happened, and it is what they do every time.

Even if they are on very distant tips of different banana leaves, separated by as much as a metre, the spiders do send their own variety of Morse code signals to each other. They use the medium of the banana plant itself as the telegraph wire!

The female spider will have deposited some female spider 'perfume' around the plant to excite the male. He is sufficiently stimulated by this to set to work with the spider telegraph.

He begins to shake his legs, something like disco dancing, and emit a series of vibrations which travel along the leaf and through the entire banana plant at a frequency of about seventy-six hertz. The female, who never for a moment moves her feet from her own leaf, answers back, in similar code, saying more or less: 'Here I am, over here!'

The male, trying to pin down this message, sends back another train of vibrations. The German scientists, in true scientific fashion, made graphs of the vibrations with an instrument like an oscilloscope. From the published graphs the signals may be closely studied, and what is most striking about them is the extremely precise timing and intervals, matched by the female's replies.

The female, by the way, unlike most human girls, never interrupts her man when he is speaking on the bush telegraph. She is pure, breathless silence. In fact, so silent and demure is she, that sometimes she forgets to answer and seems to doze off. But she answers frequently enough to give the vital information, her position.

The male finally rushes to the central stem of the plant, and cleverly – having various legs to spare – sets one of his legs on each of the descending stems of the long banana leaves. With one foot 'tuned in' to each leaf, the male is able to detect which leaf the transmissions from the female are travelling along, knows his beloved is on that leaf, and rushes to her with at least six arms open wide to embrace her. And at this point, presumably, they 'live happily ever after'. Or at least for a little while, until another telegram arrives for one of them.

Sources: Dr Jerome S. Rovner; Department of Zoology, Ohio University, Athens, Ohio.

Dr Friedrich G. Barth; Zoologisches Institut, J.W. Goethe-Universität, Germany.

MATHEMATICAL MONKEYS

A psychologist and a zoologist working together have established that chimpanzees have mathematical abilities. If you are one of those people whose face seizes up in a nervous tick whenever anyone mentions the dreaded word 'mathematics', don't start feeling inferior. The chimpanzees aren't as good as all that. But the point is, they *can* do it. They haven't progressed as far as differential equations, but they do have awareness of the concepts of number and proportion.

Using simple, easily handled objects, experiments were carried out which demonstrated clearly that the chimpanzee is capable of understanding fractions and numbers in the abstract sense and of comparing them.

Three sets of objects were prepared: round items of food such as apples, grapefruits, potatoes, all roughly the same size; circular wooden discs painted grey; and cylindrical plexiglass jars full of water dyed blue with food colouring. All three classes of objects are easy and convenient for chimps to handle and mess around with. Just the sort of thing they like doing.

Now, all these three classes of objects are totally different from one another in appearance. That is important, because the experimenters wanted them to be compared, despite their dissimilarity in appearance, according to similarities of number and proportion.

So what they did was to cut chunks out of the wooden discs of one quarter, one half, and three quarters. They filled the plexiglass jars one quarter, one half, three quarters and entirely full, with the blue water. And they cut chunks out of the apples, grapefruits and potatoes of one quarter, one half and three quarters.

They then showed all these objects to one adult chimp called Sarah and four child chimps. The child chimps weren't very bright about it, but Sarah was truly a star performer. All sorts of precautions were taken by the experimenters to make certain that no prompting was possible. When objects were set side by side, they were set there by different people screened off from one another so that neither could see what the other was placing. No names were given to the fractions of the objects. Everything was tightly controlled to make sure that what they were testing was what they wanted to test.

Sarah eagerly started taking half-discs and setting them next to half-full jars of blue water, taking apples with one quarter removed and setting them next to discs with quarter-pieces cut out of them, and so on.

She made dazzling scores showing that she understood perfectly the concepts of one quarter, one half and so forth. For she was taking completely different types of objects and setting them together when the only factor in common in each instance was a mathematical similarity!

The scientists concluded that Sarah was apparently using her head and demonstrating 'a process of reasoning akin to analogy'. They even decided that what she was doing could be expressed as follows:

x is to X as y is to Y.

(x is a piece of an apple and X is a whole apple; y is a partially full jar and Y is a full jar.)

When dealing with the fractional objects, Sarah was apparently straining her brain even further, in their view, for as they point out:

'The analogy may be considered more difficult because neither X (whole apple) nor Y (filled jar) is physically present, and therefore, must be inferred.'

The next time those mathematics professors at the universities complain that they don't have any students any more because everybody is switching to art, history and sociology, we can tell them there's plenty of talent further down the ladder of evolution. If they wait long enough, chimpanzees are bound to become entitled to admission to universities when their true merits are more widely recognised and they are freed from the oppressed status in which they are kept by homo-chauvinists.

Source: Dr Guy Woodruff and Dr David Premack; University of Pennsylvania, Philadelphia.

BIRDS WHICH ARE AFRAID TO FLY

Does a bird get vertigo? Would it get airsick in a DC-10? We tend to think of birds being 'free'. We say, 'I wish I were as free as a bird.' We dream of flying, of soaring over the earth and the wonderful feeling of exhilaration this would bring. But are birds really as happy about it as we think?

There are birds which actually suffer from fear of flying. Even the most unlikely birds imaginable are prone to this. Would you have thought that the hawk, the fierce aerial hunter who drops upon his prey like a freefall parachutist, suffers from fear of flying? Would you imagine that the aerobatically breathtaking swift, which tumbles and rolls and darts better than any jet fighter pilot, could fear the air? Or that the flapping, squawking parrot had apprehensions of any kind, much less of being airborne? All these birds have a fear of flying – over water.

Perhaps it is because birds like landmarks, or like to think that if they have to land they can set down wherever they please. Maybe they panic when they know they are committed to long flights over water and there is nothing they can do about it. But why should birds capable of dazzling flight over land suddenly seize up with terror when faced with a small expanse of water, such as a lake? You might say this was because they have some, as yet, unknown factor which actually prevents them from doing so. But it is clear that it is not due to a disability – as one zoologist has said, 'There can be no doubt that the failure to cross narrow water gaps is through choice, not inability'.

This fear of flying seems to have resulted in birds losing the ability to fly! On nearly every remote island you can think of there either are, or were, flightless birds. In many cases these have become extinct, like the dodo (which apparently was once a sort of pigeon), and the New Zealand moa. Man's invasion of remote islands, with his inevitable cats, rats and dogs, exposed flightless birds to predators with which they were unable to cope. The flightless birds were able to shed twenty-five per cent of their weight in the form of 'flight kit'- wings, flight muscles and so on. In many cases they were able to evolve into grazing animals, like sheep! Flightless birds of countless kinds are known either to exist or to have existed: flightless ducks, geese, parrots, grebes, ibises, cormorants, and even owls. These are apart from all the better known flightless birds such as ostriches, emus, rheas, and cassowaries. Birds can develop flightlessness within only a few generations. It is not necessary for geological aeons to pass while these changes are wrought in their anatomies.

Are they all cowards? Or can it be that on remote islands there are no facilities for obtaining a pilot's licence?

Source: Dr Jared M. Diamond; Professor of Physiology, University of California, Medical School, Los Angeles.

PUTTING THEIR FOOT IN IT

'How can an animal with only one foot walk on glue?'

No, this is not a riddle. It is a serious question asked by a zoologist.

Gastropods are a class of mollusc who move on one foot. They include snails, slugs, and the abalone so popular on Californian dinner tables. The problem is that with this one foot, the animal has to stick to the surface along which he is moving, and at the same time, he has to move the foot. If you had never thought about it, this seems to be a contradiction. How can the creature simultaneously stick to a surface and walk over it when he cannot lift one foot and set down another?

Obviously there is a solution, and the gastropod has found it. The gastropod foot has along its bottom surface a layer of mucus which is secreted by the creature. Anyone who has ever tried to pick up a snail will be aware that it sticks. The mucus acts as a glue. Snails crawl up walls and trees, defying gravity. They do this by sticking. But if they stick, how do they crawl? It is all very well to hang suspended in the air like a death-defying circus performer - wonderful proof of how good their glue is - but how do you both stick and slide? A sticky problem.

The secret is now known. An ingenious zoologist from Canada has done a little industrial espionage and cracked the secret formula of what we might call the Gastropod Glue Factory. Whoever concocted the gastropod mucus was a genius, no question about it. It almost defies belief. In fact, now that we know about this magic formula, we could probably make similar substances ourselves for our own use.

The gastropod mucus consists of ninety-seven per cent water and dissolved salts. As the zoologist tells us, 'The relevant mechanical properties of the material are determined by the remaining three to four per cent which is a high molecular weight glycoprotein'. So, it is only a tiny portion of the mucus which gives it its special secret. It would be too complicated to go into the details of the glycoprotein and its chemical analysis, but, basically, it expands in water and the molecules form individual crosslinks and constitute a gel network of strands. This network of crosslinks allows the mucus to have elasticity.

The layer of mucus secreted along the bottom of the gastropod foot is between .001 and .002 millimetres thick. After this was determined, mucus was gathered from slugs and studied in the laboratory at these thicknesses, along with elaborate computations of stresses and strains, as if the material were being considered for building a bridge or constructing a highway.

The gastropod moves by a series of muscular waves rippling through the bottom of its foot. With each wave, the creature moves forward about one millimetre (at least, the particular slug studied does). There are several muscular waves rippling along simultaneously at any given moment. In fact, between twelve and seventeen separate muscular waves are going on at once.

What happens is that the mucus acts as glue for small strains, but with a large strain its gel network is disrupted and it acts as a sticky liquid without glue properties. This is called 'yielding'. However, after the mucus has 'yielded', it then 'heals' when freed from the stress and goes back to the glue stage again, behaving as an elastic solid. It is thus continually alternating between liquid and solid. But it is doing this in different regions of the foot.

As each of the, up to seventeen, different muscular waves ripples along, it stresses the mucus so much that the mucus liquifies beneath the wave. Behind the wave, in the 'interwave' unstressed region, the mucus gel network is re-established, and the mucus reforms as glue again.

This means that the creature is moving along on a complex series of liquid/glue, liquid/glue, liquid/glue regions in alternation, and the mucus in its two successive states thus acts rather like a ratchet. At any moment a slug creeping along will have perhaps seventeen regions of glue on its foot separated by sixteen regions of oozing liquid. From moment to moment, the oozing liquid will freeze back into glue again and the glue will liquify, and so on and on. So instead of picking up one foot and putting down another, the creature is using one foot to serve as a succession of two different things: anchor and boat. Every time the gel network freezes into glue, an anchor is thrown down. And every time the anchor is weighed, the boat sets sail. Seventeen boats at once are going along throwing down anchors and pulling them up again without ceasing.

The slug thus manages to facilitate forward movement, while resisting backward movement. As the zoologist says, 'The result is effective adhesive locomotion'.

Source: Dr Mark Denny; Department of Zoology, University of British Columbia.

AN ANT NEVER FORGETS

Experiments have proved that ants can accomplish prodigious feats of memory and computation. They are like walking computers. The key to understanding how they think has come through studying their means of orientation and navigation as they cross what are fantastic distances for their size.

Two Swiss zoologists have performed the most ingenious set of experiments at a field station in Tunisia, using the genus of desert ants called *Cataglyphis*, which are large and long-legged ants who forage for food individually. These ants will venture as much as 300 feet from their nest in search of food. I calculate that this is like a human being walking four miles for a sandwich, and then carrying a sackful of potatoes back home for the kids.

How do the ants find their way around? Are they just familiar with their surroundings, going from tree to tree or clump to clump? No, they navigate by the sun. But in studying just how they navigate by the sun, some extraordinary discoveries were made.

When the ant sallies forth in the morning, the first thing he does is look out of the doorway and see what the zoologists call the 'landmark panorama' around the nest. This helps him get his bearings. Ants are creatures of habit and for weeks on end they emerge in just the same way, aiming towards the same landmarks.

But after moving beyond the local surroundings, an ant switches over to celestial navigation. Now he is intent on only one thing: the Puritan work ethic.

What is involved in navigating by the sun? It sounds simple, but how many people know how to do it? You have to know about the points of the compass, the sun's movement across the sky during the day and the change in the rate of the sun's movement. In the mornings and afternoons the sun is slower, seen from the earth's surface, but gives a spurt of speed at noon and fairly zips across the centre of the sky. How can ants figure all this out? After all, they are so small that even tiny errors would land them enormous distances from the nest.

The Swiss zoologists reassure us, in case we were beginning to worry about a technology explosion in the insect world. The scientists say: 'How do ants gain information about the movement of the Sun? Certainly, insects, unlike astronomers, do not perform spherical trigonometry in the sky.' That certainly puts our minds at rest.

To study the ant navigation, the zoologists painted a pattern of white grid-lines over a large area of the Tunisian desert. That was just for starters. Then they set up 'feeding stations' some distance from the various nests and managed to recruit about four hundred ants to take part in their experiments, by offering them food. When one of the subjects for experiment turned up for his free nibbles, he was grabbed and put into a moist bottle which was completely dark. No sunlight could get in at all, and the ant was kept quite comfortable for several hours twiddling his antennae in the absence of any information about the Sun's whereabouts. The poor fellow was then taken a considerable distance away and let out of the bottle. He was then tracked – not by detectives, exactly, but gumshoes of some description – and it was discovered how he managed to find his way back to his nest.

The trackers made themselves rather obvious. In fact, aluminium vehicles on wheels with long push handles, resembling nothing so much as carpet sweepers, were pushed along on top of each ant as he struggled home. These covering vehicles obscured all sight of the terrain for the ant, except for the few inches around him as he moved. He could not navigate by any landmarks at all. The only views he had were via a circular opening above him, which moved with him. He was thus allowed a good look at the sky, but nothing else. (It should be added that these particular ants do not use scent trails, and were chosen for that reason.)

Plexiglass was laid across the top opening, which was transparent to ultra-violet light, for this is important to the ants. When the Sun was screened off, and its light depolarized by a filter which stopped the ultra-violet sensors of the ant being stimulated, the ant was unable to find his way at all.

Following an ant along the desert with a carpet sweeper may seem a bit eccentric. The men doing this tracking were not told where the ant nests were for each ant, and so they were totally unbiased and could not 'lead' the ant. As long as they could get a good look at the Sun, no matter how many hours had elapsed, the ants made straight for home. Behind them went the men with the carpet sweepers, and a curious sight it must have been.

Allowing for the odd one who may have been run over by a carpet sweeper, the ants were so wonderfully successful at getting themselves out of a jam, that they caused the scientists to think again about just how this trick was done. It was proved that ants do indeed compensate for the variable rate of movement of the Sun's angular position in the sky (azimuth), even if they have not seen the sky for several hours.

Navigation by bees and other creatures has been studied and previously it was suggested that these creatures were extrapolating the Sun's position in the sky according to the time of their inner clocks. But this is too simple and is not adequate to explain how the ants did it, for they had too fine a knowledge of the Sun's movement for mere extrapolation. The zoologists explained it like this: in the main set of experiments the ants were divided into two groups: one group was trained in the morning when the Sun's azimuth moved slowly, and the other was trained at noon when the Sun's azimuth moved fast. If the ants were navigating just by extrapolation of the Sun's position, then ants released earlier should underestimate the movement of the Sun and thus deviate to the right from their home direction, and ants released later should overestimate the movement of the Sun and deviate to the left. However, no systematic deviation was observed. The ants' behaviour showed that they were compensating correctly for the movement of the Sun.

So simple extrapolation could not be the answer, for it would not allow for fine adjustments to the change in the Sun's apparent speed of movement across the sky. But this adjustment was something the ants could do! Even when released from dark bottles and in ignorance of what the Sun had been up to.

There were limits to the ants' precision, however, as a second series of experiments showed. When the ants were released after only an hour, it was found that they slightly overestimated the slow Sun movement and slightly underestimated the fast Sun movement. It was a very small error, but was a determination of 'the fine structure of the ants' internal representation of the Sun-azimuth curve'.

The zoologists finally worked out what all this means. Apparently the ants, when they emerge from their nests at various times of the day, use the familiar 'landmark panorama', in combination with their inner clocks, to note and memorise the solar positions at different times of the day. They then remember these and use them as they are needed. From all this assorted information, the ant constructs a mental picture of the Sun's movement. The zoologists put it this way: because 'the ant is able to memorise different azimuth positions of the Sun for different times of the day, it could use this information to assemble an internal model of the daily azimuth-movement of the Sun'. I don't know that I could do this without a great deal of thought. One begins to wonder just how clever these ants really are! 'Assembling an internal model' is one way of saying *the ants think and compute.*

Furthermore, experiments done with the ants in the night-time, with artificial lights, worked perfectly well also, proving that the ants 'imagine' where the Sun is, even in the dark when the Sun is behind the Earth. This shows that they have worked the whole thing out as a complete circular cycle, including the half which is invisible to them.

The next time our mathematicians are in the dark about something, perhaps we ought to turn to the ants to work it out. But the question is: do ants ever indulge in pure mathematics, or is it strictly applied mathematics? Are they simply navigators and engineers, or do they sit and speculate, and work out the value of *pi* to a hundred places? Who knows what the ant dreams about in the calm of the evening.

But the most significant thing about all these discoveries is that the ant has an incredible ability to memorise and retain information about angles and velocities. How does his tiny little brain manage all this? It poses problems for those who have theories of memory: memory traces in the human brain are said to be spiny dendrites in the brain cells which form when you try and remember something. But a spiny dendrite or two would fill any ant's brain, surely. Where is there room? How do they fit it in? What is memory? What is thought? How small can you be and still compute and memorise? The ant may be small, but like the elephant, it never forgets.

Source: Dr Rüdiger Wehner and Dr Bruno Lanfranconi; Department of Zoology, University of Zurich.

ACRES OF DIAMONDS IN SPACE

Prospectors, get your space shuttles ready! Diamond merchants, take your tranquilliser pills. The bottom is about to fall out of the diamond market. The Klondike was nothing compared to this.

Seventeen per cent of the planetary masses of the planets Uranus and Neptune is thought to consist of pure crystallised diamonds. But unfortunately for human greed, all these riches are in the outermost reaches of the solar system, and they are not on the surface (if one can call the surface of these planets a true 'surface').

How did diamonds get on Uranus and Neptune? Did God scatter them there so that our great-grandchildren could be rich? No, the lowly gas, methane, produced them. This odourless gas, which is found in the natural gas we burn, has a composition of one carbon atom and four hydrogen atoms per molecule. Scientists have calculated that in the peculiar conditions of Uranus and Neptune, the hydrogen would be stripped away and the carbon left to crystallise as a diamond sediment or residue. But much of the hydrogen would, under the intense pressures, also become solid: we would find the bizarre substance known as metallic hydrogen, which is not something you can buy in the local hardware store.

Uranus and Neptune are thought to have rocky cores, on top of which is a middle layer of what was previously called 'ice', consisting of mixed methane, ammonia and water. Then on top of that layer is a final outer surface of hydrogen and helium. Not the sort of place to go for a holiday, even if it is 'away from it all'.

Recently it has become clear from shock-wave experiments, graphs and curves, and assorted laboratory studies of that kind, that on Uranus and Neptune the ammonia and the water in the 'ice' layer would be ionized and thus separated from the methane molecules. Left to its own devices, methane, at the enormous heat of thousands of degrees and suffering from what the planetary scientists dramatically call 'shock heat', would break its bonds between its hydrogen and its carbon, allowing the carbon to become a solid residue. The intense gigantic pressures on these materials keep them all from floating off as gases and keep them liquid or solid, and, in the case of carbon, result in diamonds, just as earthly diamonds are formed under the intense pressures deep inside the Earth.

Perhaps the space epic films will now merge with the jewel thief genre, and we will have cat burglars dressed in space suits robbing Uranus and Neptune on the big screen. And, later, when travel to the outer solar system becomes as simple as flying to Honolulu, we'll all be taking our swag bags just in case.

Obviously, with all that glitter in the outer planets, diamonds are a bad long-term investment, so sell now while you still can. But, on the more serious side, our descendants may have their future changed in some more subtle ways by the presence of so much 'hard stuff' so far out from the Sun. Because diamonds, renowned for being the hardest substance on Earth, if mined in vast quantities in the outer solar system, long in the future may have a unique engineering role when utilised in possible futuristic construction schemes. Not to mention the possibility of finding numerous cutting tools for the roughest work, on the spot.

We may be able to mine the diamonds by automatic robot scoops which drop down and emerge with mouthfuls of gleaming 'fish' from beneath the surface seas. The metallic hydrogen would presumably melt on the way up, and spill away. Then we could use the diamonds to form walls and floors of cities and colonies in deep space, saving the transport from Earth of proper metals. They would be extremely safe against meteorite bombardment, made of such hard material. And what the buildings lacked in sunlight they would make up for in pure glitter. 'Rhinestone Heaven', but made of the real thing. The gaudiest space cities anybody ever heard of. Superman would feel right at home. Of course, the survival of automatic robot probes at those intense temperatures and pressures is probably not possible, so some other way of getting at the material would have to be devised. Maybe it could be blasted up by laser beams and caught before it falls back again. But you can be sure that all the world's engagement rings are not going to retain their value forever. When diamonds beckon, man, bewitched, follows. What a boost this will give to the space programme.

Source: Dr Marvin Ross; Lawrence Livermore National Laboratory, University of California, Livermore, California.

LEFT - HANDED AND RIGHT - HANDED PLANTS

The duckweed, a tiny floating plant of still water ponds and ditches, is a favourite food of ducks. They scoop it up greedily, though it can still spread to such an extent that it will sometimes cover the surface of a pond like a green blanket. Under this blanket, in stagnant water, are often large quantities of insects, also tasty morsels for feeding birds.

The tiny duckweed is not as simple as it seems. It is left-handed or right-handed, just like people. Only in recent years have the full facts about this little plant become known, because the duckweed flowers very, very rarely. It took the most intensive study of induced growth in laboratory conditions, over a considerable period of time, to uncover the secrets of the duckweed's handedness.

It had been noticed for some time that new shoots of duckweed grow out of two particular areas, called 'pockets', on a mother frond, and that they do so in a peculiar fashion showing handedness. It used to be thought that the duckweed was left-handed, and always produced its first baby shoot from the left pocket. Then it produced a shoot from the right pocket, then returned to the left again, and then back to the right again, in a back and forth sequence.

Now it has been discovered that the duckweed is not left-handed after all, but right-handed. The rarity of the flowering in natural conditions had obscured this fact. It is really the flowering potential which takes precedence over the shoots. Because the flowers normally appear from the same right pocket as the second shoot, and flowering takes priority over shooting, the duckweed is forced to shoot on the left, because the right pocket is reserved for a flower which rarely appears. After the first shoot on the left, the non-appearance of the flower on the right allows the second, right-hand, shoot to appear and the sequence continues back and forth. So, in fact, the duckweed is actually right-handed, but masquerades as left-handed.

But this is not the end of the story. Under the most prolonged laboratory-induced conditions of growth, the duckweed will eventually produce a double-flowering variety which flowers on both sides. Now what happens to the handedness? This is where things really get interesting.

From double-flowering plants, shoots emerging from opposite sides give rise to 'lineages' or families of descendants with opposite handedness. And they reverse the normal pattern of single-flowering plants. For whereas babies from single-flowering normal plants have the same handedness as their mothers, babies from double-flowering mothers have the opposite handedness to the particular pocket from which they grew, and also the opposite handedness to each other, if from opposite pockets. Are you confused? So are the ducks.

Source: Dr P. Doss; Ornamental Plants Research Laboratory, U.S. Department of Agriculture, Puyallup.

THE PLANT THAT LAYS EGGS

The passion flower is appropriately named. It doesn't kiss and cuddle. But it does lay eggs. And, what is more, like the proverbial goose, it lays golden ones.

Or perhaps we should look at it this way: the ancient Chinese philosopher Chuang-Tzu made the famous statement, 'Last night I dreamt that I was a butterfly. But now that I am awake, how do I know that I am not really a butterfly that is dreaming he is a man?'

The passion flower dreams that it is a butterfly.

Probably you are wondering how I can reconcile all these statements. The answer is as follows. The passion flower pretends to be a butterfly of the neotropics call *Heliconius,* which lays golden eggs. *Heliconius* is a ferocious predator of the passion flower. The passion flower has spent millions of years trying to outwit various insects - by exuding poisons or smelly substances - and *Heliconius* is the last, but by no means the least, predator left who really goes for the plant. One single larva can defoliate an entire passion flower plant. That is some appetite!

The passion flower must have racked its - well, perhaps not its brains - to try and think up some gimmick to deal with this terrible, voracious butterfly. And it came up with a brilliant idea. It had noticed that the *Heliconius* is a very fussy egg-layer. The butterflies tap and stomp on the leaves, fuss about, probe and poke and sniff, and behave a bit like a male hairdresser beautifying one of his clients. They only want to lay their precious eggs on the very best and tastiest bits. But that is not their only consideration. They are worried about other eggs being nearby.

And for that there is a good reason! The butterfly larvae are fiercely carnivorous, indeed, not to put too fine a point on it, they are cannibals. A *Heliconius* larva would as soon gobble up its pal in the next egg as spit.

Mama knows about baby's little blemish to his otherwise spotless character. (What's a little murder in the family, if it's all kept quiet?) And so she tactfully lays her eggs a sufficient distance from other eggs so that junior will not be tempted to buttercide. But, of course, there is the even more crucial factor: one wouldn't want junior to get eaten by one of those next-door larvae.

The passion flower, meanwhile, the world's greatest expert on *Heliconius*, believes in the maxim: 'know your enemy'. It has seen these fussy predators flutter and strut enough to know their weakness. And, like a true genius of the plant world, it formulated a grand plan: it would fool the butterflies into thinking there were already eggs laid all over its tendrils' tips, the juiciest and most sought after places, and then the butterflies would be apprehensive and not lay the eggs there.

This is exactly what happens. The passion flower actually grows perfect replicas of the *Heliconius* eggs, coloured just the right golden colour. It lays its own eggs, as it were, to forestall the real ones being laid there. And the butterflies fall for it. They seem to recognise their own eggs mostly by sight rather than smell, and when they see a fake egg they flutter off and seek somewhere else. This has all been meticulously studied and proved by two scientists with long, involved experiments. The discoveries only became known in 1981, and constitute a striking illustration of, 'a plant structural trait resulting from co-evolution with an insect herbivore'. The implications for evolution theory are quite considerable. It is only these rare plants, such as the passion flower, with only one insect predator, which offer the opportunity to study such 'co-evolutionary' developments.

Not all species of passion flower grow eggs. But the ones which do seem to have the capacity to expand their range and thus profit from their brilliant innovation. The scientific studies have unequivocally established that the butterflies really do flutter away without laying as many eggs as they do on passion flowers without the fake eggs. The trick works. And it is the colouring which is all-important. The scientists painted green dye over some of the fake eggs and the butterflies didn't take any notice of them anymore. The passion flower actually had to go to the trouble of finding the chemical ingredients to lay pigmented eggs which match almost exactly the colour of real butterfly eggs which are on the point of hatching.

The studies have shown that passion flowers all over the world seem to be catching on to this idea. As the scientists conclude, 'The evolution of egg mimicry is an ongoing process'. It is spreading. One day, who knows, perhaps they will even try to fry them.

Sources: Dr Kathy S. Williams; Department of Biological Sciences, Stanford University.

Dr Lawrence E. Gilbert; Department of Zoology, University of Texas.

THE HERB THAT MAKES INSECTS STAY YOUNG

Imagine going to the supermarket to buy food for your children, which they find absolutely delicious. But this food has a strange effect upon them: it stops them growing up. The eight-year-old girl stays eight, or perhaps creeps up to nine. The ten-year-old boy will never shave because he never passes puberty. Imagine all the children in the neighbourhood staying children, and all rushing every day to the supermarket for this delicious food.

But what happens? The years go by, you get shaky on your legs and all your friends are dying. The older generation are disappearing from the streets. And still these eight, ten, twelve year-old children are walking about without the slightest desire to marry or have children of their own. They may be thirty or forty years old, but they have not grown up, are immature and still behave like kids.

This is what happens to insects which suck the delicious juices of the herb, basil. They never grow up. Basil actually manufactures insect hormones which have this effect on them.

Basil is one of the most delicious herbs for cooking. It is the best flavouring for tomatoes, and no self-respecting French chef would be without it. In fact, a secret of the Parisian chefs is that they use distilled oil of basil to flavour their tomato soups and sauces. For it is not always practical or possible to get the chopped-up fresh leaves, especially when cooking for large numbers at once. It is this very oil of basil, exuded by the crushed glands on the leaves when the herb is handled by man, and present in the juices sucked up by the insects, which contains the juvenile hormones. The distilled oil is simply a highly concentrated form of it, one or two drops representing the essential oils of an entire plant.

Studying the concentrated oil, scientists have analysed its constituents. The Latin name for basil is *Ocymum*, so one of the crucial ingredients is a compound which has been named *ocimene*. The two insect hormones have *ocimene* in them and since they keep the insects juvenile, they have been named *juvocimene* 1 and *juvocimene* 2.

These discoveries came as something of a surprise to plant and insect specialists. It was well known that a variety of plants manufacture poisons and other chemicals to kill, repel, or resist insects feeding upon them. But it is only in recent years that it has been found that plants actually manufacture insect hormones, as a far more subtle and sophisticated means of defence.

The insect eggs hatch into nymphs and they then grow through various stages until they can become proper adults, capable of reproduction. But the *juvocimene* hormones of basil are so powerful that they accomplish what the scientists dramatically call 'hormonal derangement' of the insects' development. It makes the insects either stay nymphs, or stop their growth at 'nymphal-adult intermediate' stage. The nymphs of many insects are characteristically sluggish, inactive and disinclined to feed. Therefore, it looks as if basil has killed two insects with one stone: stopping the insects from feeding and stopping them from reproducing. A devious double strategem.

Strange to say, for thousands of years basil has had a reputation of having an affinity with scorpions. Perhaps experiments should be done studying the effect of the insect hormones on them. Scorpions are said to like sitting under basil bushes. Are they getting high on a hormone, or do they just have discriminating senses of smell and a flair for haute cuisine?

Does basil contain an elixir of immortality which operates on more than just insects? The Emperor Julian's doctor believed that basil would protect men against all poisonous bites of insects. Modern herbalists sometimes maintain that basil repels flies or mosquitoes if left about in the house, or if the oil is rubbed on the skin. It is interesting that insects and sex have traditionally been associated with this herb which effects the sex life of insects. Reputed to be a powerful aphrodisiac, especially for horses, the famous seventeenth century herbalist, Nicholas Culpeper, wrote of basil in cryptic vein: 'It expelleth both birth and after-birth; and as it helps the deficiency of Venus in one kind, so it spoils all her actions in another. I dare write no more of it.' Perhaps he had heard about the insects.

Source: Drs W.S. Bower and R. Nishida; Department of Entomology, New York State Agricultural Experiment Station, Geneva, New York.

THE BUTTERFLY THAT CAN TELL MEXICAN MARIJUANA FROM TURKISH MARIJUANA (WHICH MAN CAN'T)

Question: What is the connection between cabbage patches and drug smugglers?

Answer: The much-dreaded Large White Butterfly.

The Large White (*Pieris brassicae*) is known to all vegetable gardeners as one of the major pests he has to combat. This Cabbage White, as it is sometimes called, with its companion the Small White, lays vast quantities of eggs on the cabbage plants. From these, masses of squirming caterpillars appear, which devour the cabbage plants with ferocious speed.

Two ingenious scientists decided to spray cabbage leaves under glass with soupy extracts of the juices from two different kinds of marijuana, *Cannabis sativa*, CBD (from Turkey) and THC (from Mexico). These two forms of marijuana contain nearly indistinguishable variations of a substance called a cannabinoid (named after Cannabis). These two cannabinoids are distinguished only by a single ring closure in the molecular structure. To a human the two substances taste and smell identical.

Having sprayed the cabbage leaves with the two similar substances, the scientists discovered that the Large White butterflies which were trying to lay their eggs on the cabbage leaves were totally repelled by the Mexican THC marijuana spray but only mildly repelled

The Butterfly That Can Tell Mexican Marijuana From Turkish Marijuana

by Turkish CBD marijuana.

There were also some cabbage leaves sprayed with ordinary water. Egg-laying on these leaves exceeded egg-laying on the leaves sprayed with Mexican marijuana juice by twenty-two to one, whereas it exceeded egg-laying on the leaves sprayed with Turkish marijuana juice by only three to one.

The Large White butterfly, therefore, has an extreme aversion to Mexican marijuana but only a mild aversion to Turkish marijuana. Considering that we can't tell the difference, the butterfly is therefore extraordinarily sensitive.

Drug smugglers may live to rue the day that this was discovered. Police investigating marijuana shipments may need to establish scientifically where the marijuana came from. Is it Turkish or is it Mexican? The Large White butterfly, pest of the cabbage patch, may end up sending people to jail, and may expose their sources of supply and help trace the smuggling routes.

The next time you are travelling and a group of special policemen come along trying to sniff out drugs in the luggage with animal help, it may not be a dog that is on the end of the leash. The guilty drug-smuggler who sees a flutter of white wings on the end of a leash, coming towards him along the train carriage, or waiting at the airport customs barrier, had better be ready to run for it.

Or will marijuana-smugglers begin to carry supplies of fresh cabbage leaves in their pockets when travelling, just in case?

Source: Miriam Rothschild, Ashton Wold, England and J.W. Fairbairn; School of Pharmacy, University of London.

'YOU LOOK GOOD ENOUGH TO EAT'

Imagine sixty sex-starved men and women confined to a dark cavern, deep in the Earth. They all have chronic laryngitis and cannot speak. Each one of them is given a lantern with a sliding shutter with which he and she can signal along the dark passages of the cavern. The men and women are divided into three groups – the Greens, the Reds, and the Blues. The Greens are told they can make two flashes with the lantern and will get two flashes in reply; the Reds are told to flash three times, and the Blues four times.

By answering the differently-coded flashes, each man or woman can get a reply from another member of his or her group. If a member of the opposite sex replies, they can answer each other's flashes as they get closer, until they finally meet and, in a mad frenzy of passion, copulate. All of the people in the cavern are sex maniacs, and this is what they are interested in above all else.

After a time, however, the females of the Greens develop a very strange and perverse desire: hunger, in the form of an overwhelming desire to eat the men of the Blues and the Reds. This cannibalistic urge overwhelms the previously dominant sex urge. The males of the Greens become extremely upset and hit upon a devious strategem to entrap their women into continuing to mate with them. The female Greens have been simulating the flashes of the Blues and the Reds in order to lure them to them and eat the males. Horrible cries are heard throughout the cavern as Blue and Red males are torn to shreds and devoured by the terrifying females of the Greens.

The male Greens decide to simulate the flashes of the Reds and the Blues themselves, but not to lure the Reds and Blues to them; instead, they want to fool their own women and make the female Greens think they are males whom they can eat. It is a risky business, as it means they might also be eaten; but their sex drive utterly masters their reason, and they take the chance and hope that some group cannibalism taboo will save them from their predatory women.

It works! The women of the Greens, thinking they have found tasty morsels, are lured to their own scheming menfolk, and then are pounced upon and raped. After the coital act, the occasional Green male is eaten as dessert, but most of them survive. Thus, a pattern of behaviour is set up which continues, in a thrilling atmosphere of danger, throughout all the successive generations of men and women living in the dark cavern in the decades to come. A nightmare? A fantasy of violence and horror? It is real life, and entirely true: but not for human beings, thank goodness. It describes the world of *fireflies*.

If you have ever wondered what it is like to be a firefly, now you know. They never think of anything but sex and food, and cannibalism is rife. They live in a terrifying, dangerous world full of crunching carcasses and screams and groans, interspersed with passionate copulations.

Only towards the end of 1980 did the thirteen years of research by a Florida entomologist give the full picture of firefly activities. The Greens of our cavern fable are a species of firefly called *Photuris*. The Blues and the Reds are other victim-species.

The *Photuris* have their own signalling system of flashes, high up near the tops of pine trees. The males make a specific flashing pattern and fly towards answering flashes, eventually reaching the firefly who answers, and in most cases this being a female, they mate. But like the cannibal-women of the Greens, the *Photuris* females 'in addition to using flashed signals in their own sexual communication, mimic females of other species, attract males, and eat them' in the words of the entomologist. This is what the scientist calls 'aggressive mimicry'.

The victim-species of fireflies live lower down, less than three metres above the ground in grassland and amongst pine trees. The scientists who study them call this their 'flight space'. The normal flash of the *Photuris* fireflies to each other is a slow flash-phrase of three to seven flashes. But the cannibalistic female *Photuris* mimic the codes of the victims in the lower flight space, ranging from a series of short flickers to minutes-long bright sustained glows. Scientists studying the flashes use recording equipment in the field which is a photomultiplier magnetic tape system. From these recordings they obtain line-graphs which are easily set side by side and compared. It may readily be seen by this method that the flashes of the victim-species and the mimicked flashes of the predator *Photuris* females are indistinguishable.

The *Photuris* males, perceiving that their females are luring male victims from the lower regions whom they eat, pretend to be those victims by themselves mimicking the victim-males' responsive flashes. This is like a con-man conning a con-man with his own con. And it is very successful. The *Photuris* females, expecting a good meal, through their own deceit, are lured by the counter-deceit of their own males to unexpected sexual rendezvous. As the entomologist suggests, it could be concluded that by this means 'hunting females (*femmes fatales*) that are sexually unresponsive may be located and raped'. But his colourful language does not stop there. He suggests that the males take a big risk in doing this, as they are smaller than the females and may themselves be eaten. Their sex-drive being so overwhelming, they may decide that for the sake of one more fling with a female they are prepared to die. The scientist calls this 'kamikaze-copulation', which fortunately is not in fashion with humans yet. And as he describes it: 'thus, old males with low probability for survival to the next evening's mating flight might feasibly advertise their candidacy for cannibalism by a mate, this being the ultimate in nuptial feeding'.

In others words, the male fireflies may be saying: 'I may be tricking you into thinking I'm one of the more delicious ones from the lower flight-space, but I really am good enough to eat. Try me. But just before you take the first bite, let me just ... as it were, embrace you ... if you see what I mean'.

The incredible cruelty and violence of the insect world is brought home by these heartless events. Food and sex, in monotonous regularity, as the sole aims, dominate this dread horizon. Fireflies are so desperate for sex that they are driven to the uttermost extremes of intelligent devices of which their tiny insect brains are presumably capable. But no higher aspirations seem to enter into it. The complicated nature of their mimicry behaviour is truly astounding, and the entomologist says of it: 'such a complex of behavioural mimicries is without known parallel in the animal kingdom'.

So despite the fact that the fireflies have evolved a system of impersonation and deceit more complex than any known even among the higher animals, excepting man himself, the ends towards which it is all turned mock the apparent intelligence involved in creating it. It is as if a nymphomaniac had worked out Einstein's theory of relativity simply in order to entice an attractive physicist into her bed. So low do the fireflies stoop, who soar so high. And those attractive flashing lights on a warm evening in the more temperate climates are but the sad beacons of aimless passion and sudden death.

Source: Dr James E. Lloyd; University of Florida at Gainesville.

THE WIND IN THE LILIES

Yellow water lilies actually gasp in huge amounts of air, several gallons per day apiece. These heavy-breathers may look peaceful as they lie on the surface of a pond, but in reality they are gasping for breath.

Plant roots need oxygen, and the roots and rhizomes of the yellow water lily lie buried in the muddy sediment at the bottom of the ponds, where no air can reach them. But it was only in 1980 that it was discovered how desperate the lilies were for air. They suck in air and force it down through themselves at the incredible rate of ten feet per minute!

It is primarily the young leaves that do the heavy breathing. You can see it at work by taking a transparent jar and placing it over the young healthy leaf of a water lily in the sunlight. After some time, the water level inside the jar will rise as much as five inches above the surface of the pond, filling the vacuum caused by the leaf pumping air down through its pores into the root system.

The water lilies don't take any particular element from the air, they suck the air as it is, though it is the oxygen they are after. As they suck down the air at a rapid rate through the young leaves, the older leaves emit bad breath. They are outlets for stale old gas full of carbon dioxide. Therefore, breathing in through the young leaves and out through the old leaves, the yellow water lily is a completely ventilated plant with a constant and rapid flow of gas passing through it.

What quantities of air are breathed by the water lily? A

single plant in one day can actually pump as much as twenty-two litres of air down to the roots. The air and gas don't stand still, but are continually being circulated through the entire plant. The pores of the water lily therefore, in the words of the scientist who discovered it, 'behave as a pressurised flow-through system'. If the root through this system takes in twenty-two litres of air a day, from this quantity of air it will extract four point six litres of oxygen. The gas expelled by the plants is not just carbon dioxide, but also significant quantities of methane. The energy for the pumping comes from the sun's heat. When the leaves are exposed to sunlight, the interiors of the leaves are warmer than the surrounding air, leading to air rushing in and also to the vapour pressure of the water in the leaves rising above that of the surrounding air. The pores of the young leaves which suck in the air are linked to the pores of the old leaves, which vent the vapour pressure of the young leaves. This powers the breathing in as well as the breathing out. All this, and no lungs.

Source: Dr John W.H. Dacey; Michigan State University.

DRINK SHRINKS THE BRAIN

Alcoholism is a very serious disease, and many of its effects are well known to everyone. But one of the most recent discoveries about alcoholism is perhaps the strangest of all. It has been shown that it shrinks the brain.

The human brain is divided into two primary parts, the left hemisphere and the right hemisphere. Alcoholic brain-shrinkage takes place only on the left side, so that only the left hemisphere, or half of the brain, actually shrinks. The right half of the brain seems to remain more or less the same.

In order to determine the facts about this, scientific experiments were carried out with carefully selected groups of alcoholics. They were all chosen below the age of forty-five, to eliminate possibilities of brain changes due to age. None with a history of brain injury were included, none who had dropped out of school were allowed to take part in case it indicated early brain damage. A mixed group of men and women was studied, whose average age was about twenty-nine. They had all been drinking for over eight years on average. A similar group of non-alcoholics was studied for comparison.

Each one of the people in the two groups was then given three X-ray scans of the brain, at three different depths. The results of the X-rays showed that the alcoholics had, on average, twenty per cent shrinkage on the left half of the brain. That means that one-fifth of the left hemisphere had gone.

It has long been suggested that the demon drink steals our wits away. But now we know which ones and how much of them. If you are a heavy drinker and begin to feel light-headed after a few gin and tonics, make sure you don't just feel light-headed on the left side. For then it's time to turn to orange juice.

Source: Drs Charles Golde, Benjamin Graber, Irwin Blose, Richard Berg, Jeffrey Coffman and Solomon Bloch; University of Nebraska Medical Center.

STORMS BENEATH THE SEA

It used to be thought that the bottoms of the world's oceans were quiet, tranquil, somnolent places where the rough and tumble of the waves and currents never stirred the dreaming worms that burrowed quietly in the mud. But in 1979 and 1980, surprise findings meant that this view had to be abandoned.

Strange and, for the moment, inexplicable storms have been discovered in a region of the western Atlantic off the coast of Nova Scotia, reaching down the coast of New England. But these storms take place at the bizarre depths of 15,000 feet. That means they occur as far below the surface of the ocean as the Andes Mountains rise above sea level.

Scientists who study the ocean floor and ocean currents have various means of inspecting events at these depths. They can send down bottles and bring up samples of the water to see how much mud it contains. To their astonishment, when they did this in the area off Nova Scotia, the samples did not contain the usual crystal-clear water like spring water which one gets from the ocean bottom. Instead, the samples contained two hundred times more mud particles than usual, as much mud as you might find in a great muddy river like the Amazon or the Nile.

Currents were flowing at these great depths at rates of eight to ten feet per second, which is much faster than many rivers. The currents seem to be flowing generally to the south-west in the deep ocean off the continental shelf.

Photographs were taken of the sea bottom at these depths. Some photos showed completely undisturbed areas of mud with worm-burrowings which had obviously not experienced the slightest ripple of water over them. Other photos showed that the sea bottom had been partly smoothed over by currents, leaving only the topmost ridges of the burrowings showing. And finally, photos showed completely smoothed sea-bed regions where violent currents had rendered the sea-bed completely featureless and eliminated all traces of the worm holes.

But the most dramatic evidence of the violent undersea storms was obtained by devices called transmissometers. These include little lights and instruments to read how much light can be transmitted through tiny samples of the water. This is a way of judging how murky the water is without having to haul up bottles the entire 15,000 feet each time.

Surges of storm current were discovered lasting only three or four days. One reading showed the 'transmissibility' of the water to light was only 0.01 per cent. (Clean sea-water normally has a value of sixty-eight per cent.) In other words, even the muddiest pool in the dirtiest track you have ever seen would do well to exceed these incredible values. Scientists found changes of twenty per cent in a single hour, and five per cent in one minute. This means that clouds of mud are being whipped up which are only a few hundred yards long, and which are hurtling along the sea bottom very much like atmospheric storms.

The violence of the undersea activity off Nova Scotia seems to be a deep undersea equivalent of the famed Gulf Stream, which crosses the Atlantic at roughly the same speeds. But the Gulf Stream is largely a surface phenomenon, and does not extend to great depths. Although it is true that curious deep sea sloshing movements of water take place beneath the Gulf Stream off Cape Hatteras in North Carolina, reversing direction every month or so and reaching speeds of twelve feet per second, these are thought to be linked, in some yet undiscovered way, to the Gulf Stream itself. But the undersea storms off Nova Scotia are not near the Gulf Stream and must have some mysterious cause quite of their own. Monsters of the deep waggling their tails, perhaps? If so, they would need to be larger than ten ocean liners, all sitting in a row munching on seaweed, with severe indigestion. It is more likely that this is a case of a current-lashed, wild seabed which suffers as much turbulence as the coast of Cape Horn 15,000 feet beneath the surface of the waves – a mysterious storm centre of the world's oceans, home of veritable undersea hurricanes. This is an enigma which still awaits a proper explanation.

Sources: Dr Ronald Zaneveld; Oregon State University. Drs Charles Hollister, Albert Williams; Woods Hole Oceanographic Institute.

Dr George Weatherly; Florida State University.

JUMBO JETS FOR HONEYBEES

Bee swarms, which may consist of more than 30,000 individual honeybees, flying together in search of a new nest, act like individual animals to an uncanny degree. These swarms are highly structured entities which accomplish fantastic feats.

If the temperature of the surrounding air is 1^{0}C, just above the freezing point for water, the swarm generates and regulates its internal heat and can achieve a 'core temperature' at the centre as high as 46^{0}C!

The bees forming the outer wall of the swarm can crawl in and get warm when they start to feel cold. The swarms are not just jumbled masses but are structured in layers, with corridors along which bees can crawl in flight. The swarm is therefore rather like a well-heated and efficient jumbo jet which can carry over 30,000 people. The bees have therefore made the greatest advances in mass air-travel, and they don't even have to build the planes. For they themselves make the walls and corridors of the transport craft, without the use of a single piece of material or any machinery. They are the airplane.

Bees inside the swarms suspend themselves from each other in curtains, remaining motionless, to form the walls of the very spacious jumbo jet corridors inside the moving entity. Many bees can walk along the corridors side by side. When things get rough, and the outside air becomes extremely cold, the corridors are entirely packed with inert bees, who bunch together in order to stop up the corridors to prevent heat loss by convection from inside the swarm to the surrounding air. All this, of course, is taking place while the thousands and thousands of bees are in mid-flight.

Studies of bee swarms were never properly done until 1981, and this information about them came as a surprise. The way in which the swarm regulates its temperature and shape and structure resembles the way in which a vertebrate animal does. Why do bees swarm? The reason is that bee hives can become overcrowded and an old queen and about half the workers of a hive decide they need to look for another place to set up house. So they decide to fly around and scout out the property market.

The new house which this clan need to find has got to be the right sort. Faulty heating will finish them off. They need a nice hollow tree or something simple, but it has got to have good insulation. A bad choice will mean the extinction of the colony, so there can be no mistakes.

The thousands of workers begin to get themselves ready. They queue up for an enormous meal and gorge themselves shamelessly on – what else? – honey. But then, as they have spent all their lives gathering and making the stuff, who can criticise them for raiding the pantry, especially when it could be their last meal?

Stuffed full of quick energy, the bees get ready. They have got to find the new home before they lose their strength from this honey feast. Time is of the essence. They gather themselves together and get organised. Arousal and take-off can only occur when a completely uniform temperature is obtained throughout their massive ranks. Around the swarm there forms a roughly spherical mass of 'outsiders' who have the most difficult job. These outsiders form what is called the swarm mantle. If an intruder approaches, these guardians of the swarm raise their abdomens and expose their stingers. In cases of serious trouble, a second phalanx of warrior-bees are at the ready at all times. They are waiting inside the core of the swarm but at a signal rush along the corridors and emerge, squeezing through the outside mantle, and attack any enemy. They are the 'palace guard'.

Meanwhile, the other bees are arranging themselves in rows to form the walls of the internal corridors - living partitions.

The swarm adjusts its shape and size, elongating or compressing. If things are too chilly, the outsiders in the mantle all shiver like crazy, in order to generate extra heat. If some of them feel really under the weather, they dive into the corridors where they warm up to about 30°C or more, before returning to duty. There are always other fellows ready to take their places while they are on leave.

At last things are right, an even temperature has been generated all through the swarm. Take-off! Away they go, veering around, and start heading off in whatever direction they seem to think holds promise.

With any luck, the bees find themselves a suitable tree-trunk. But when will we achieve this breakthrough - jumbo jets without any spare parts? Air transport made entirely of the passengers themselves!

Source: Dr Bernd Heinrich; University of Vermont.

THE TREE THAT COMMITS SUICIDE

Mother-love with a difference is shown by the huge tropical tree that dies for its young. *Tachigalia versicolor*, which has only a Latin name, grows over 120 feet high. Year in and year out it saves energy by not flowering, while it makes its way slowly up to the top of the jungle canopy. Eventually its large, spreading branches overtop its neighbours and it stands graceful and magnificent, a giant among other giants.

Then one day without warning, the tree begins to flower. This may be at any time of year, since it recognises no flowering season for its one big event. Its flowering is a dramatic explosion, as a single tree may cover its top in seven thousand blooms. The vast number of flowers turn into green leaf-like fruits six inches long, which slowly turn brown and are dispersed by the wind. All round the tree in its jungle home in Panama, Costa Rica, Columbia, or the Amazon, these thousands of seeds fall and many of them take root and grow up as seedlings.

But a seedling stands little chance in the highly competitive and ruthless environment of the tropics. Struggling away at the bottom of the forest, nearly lost in darkness and choked with vines and other plants, the seedlings of the giant tree threaten to become a lost cause. How can these tiny seedlings struggle upwards to the magnificent 120 foot height of the parent? How can *Tachigalia* multiply and survive?

At this point the great tree makes what appears to be a unique gesture in the entire plant world – it gives its life for its young. In order to make space for its seedlings to grow and give them the access to the sunlight that they need, the *Tachigalia* commits suicide. Within months of sheddings its seed in its single flowering splendour, the tree dies. Then one day, in a strong wind, its huge rotting trunk crashes through the tree and vine canopy, ripping an enormous hole 120 feet long in the jungle roof. Many side rents appear, as neighbouring trees of different species are felled by the crash.

As wind-borne seeds from trees rarely fall more than 200 feet away from their parent, by far the majority of the seedlings of the *Tachigalia* are within range of the falling trunk. Some are utterly crushed, but large numbers are given access to sunlight and space and grow and expand. They now have the crucial 'light gap' in the canopy which all plants need in the jungle. They begin their slow climb upwards, and one will make it.

The suicidal behaviour of this tree was not known until 1967, when it was first reported by a research station of the Smithsonian Tropical Research Institute in Panama. In the years after that, a search was made for the trees, and nearly 500 were spotted and tagged. In 1974, the peak year for tree-suicide, eighty-five trees killed themselves for their young. This amounted to nearly one-fifth of the known numbers of the trees in Panama. In other years, sometimes none died at all.

The only escape from death by the tree, once it flowers, is if it restricts its flowering to just a few branches, those branches then die but the tree itself can live on. However, only three trees have been that clever, out of the 500 studied. The rest have either not yet flowered or have gone the way of hari-kiri. There has also been no shooting or sprouting from the bases of the trunks, or from any of the root systems of the old trees. They do seem to die utterly and completely. The scientist who reported these facts was in no doubt of what the trees were doing – he called it a 'suicidal act of reproduction'.

But the trees live on in their children. Is this the most extreme form of mother-love in the plant world?

Source: Dr Robin B. Foster; University of Chicago.

BABIES WHO ARE BORN ANGRY

In the spring of 1981 it was finally demonstrated that human beings are more or less aggressive by nature depending on the amount of sex hormones to which they were exposed in the womb. But this points to a very serious problem. In the good old days, a baby took its chance according to what sort of mother it had and whether she had lots of sex hormones in her blood or not. But that is no longer the main concern. The fact now is that millions of babies have been apparently made more aggressive for life because their mothers have taken certain hormones during difficult pregnancies. There can hardly be any other result than that society itself has become, and will become, much more violent as a direct consequence of this innocent and harmless treatment.

Over the past thirty years, certain hormones have been given to expectant mothers if the women were in danger of miscarriage or abortions. The specific hormones were progestins and oestrogens. For many years it has been established that orally administered synthetic progestins taken for this purpose have had the result of causing the genitals of female babies to be 'masculinised' in shape, which in extreme cases can be a very embarrassing problem. But now the findings about aggression far outweigh the findings about sex in their importance. Sex and aggression are of course linked, and both are influenced by hormones. But whereas the 'masculinisation' of female genitals only affected eighteen per cent of the babies of women taking the hormones, the effects on aggression seem to be much more widespread.

It has now been proved that 'angry babies' have been born in great numbers and that they have been growing up into angry men and women. People who maintain that society is getting more aggressive and violent may indeed be right. The future prospects of every one of us could be affected and it is therefore not just a theoretical problem.

What makes the problem so staggering is that millions of women, and even more millions of babies, are involved. But saying 'babies' is misleading: many of these babies are now in their late twenties or early thirties. They constitute, in fact, the entire new generation. What we have done without realising it is to breed a generation of far greater violence than any which has existed since, presumably, the time of the cavemen.

How certain can we be that this alarming state of affairs really exists? The studies which reveal it have not been done on rats or other laboratory animals. The studies were done on actual human beings, alive today, and aggressive today.

A large number of children aged, on average, about eleven and a half years were studied. Those who had been exposed to the sex hormones during pregnancy were compared with their brothers and sisters born earlier, who had not been exposed to the same drugs. The lowest age examined was age six, the eldest was eighteen years old. All of the hormone-exposed children had been exposed to the drugs before the seventh week of pregnancy. The duration of the hormone drug treatment ranged from four and a half weeks to as much as thirty-one weeks. Every subject examined was therefore exposed to the hormones during the crucial period in the development of the embryo when the shape of the genitals is determined, which is the period when aggressive susceptibilities also seem to be highest.

Intelligence tests were given to the children, but no differences were noted in that area. The difference arose in the aggression tests, which are called the Leifer-Roberts Response Hierarchy. The test requires a paper-and-pencil instrument which assesses the potential for aggressive behaviour in people faced with imaginary conflict situations suggested during the test. As expected, the males all scored higher for aggression than females, but that is common with all classes of mammal, including man. The children who had been exposed to the hormones in the womb had drastically higher scores on the aggression tests.

The girls' increase in aggression was not quite as bad as the boys'; they were one-third more aggressive. But the boys were almost twice as aggressive. It was also discovered that one particular hormone was the worst of all, and children exposed to it in the womb were by far the most aggressive. The dangerous hormone is known as 19-NET for short, its full name perhaps also being worthy of recording here: 19-nor-17alpha-ethynyltestosterone. This is the same hormone known to have the most drastic effect on 'masculinising' the female genitals.

Although the girls suffered less from increased aggression, their behaviour seems to have been significantly 'masculinised'. They were not found to have been tomboys when young so much as to have become prone to violent and angry fits upon reaching adolescence, at which time they also began to act in a more masculine manner. Behavioural changes in the boys were obvious at a far earlier age, however. Their increased aggression showed up as small children and they had far more need of serious discipline early in their school careers.

It is certainly strange that dosages of hormones at a crucial stage in the embryo can affect a person throughout his or her life, but it is now known to be true. It would seem advisable at least to drop the use of the hormone 19-NET, if not to review the entire range of hormones used in the difficult pregnancies. But we cannot undo thirty years of damage. We have made the world far more violent and dangerous perhaps than it ever was.

Source: Dr June Reinisch; Rutgers University.

You don't have to be a Nobel Prize winner to read and use SCIENCE IN EVERYDAY LIFE. Drawing on an astounding range of facts, figures and information, this fascinating book lucidly answers the thousand and one questions we ask about the world every day . . .

How are temperatures converted from Fahrenheit to Centigrade?

Does lightning travel upwards or downwards?

Can gold be extracted from sea water?

When was the needle invented?

Where did air come from?

SCIENCE IN EVERYDAY LIFE

Straightforward answers to all those difficult questions.

SCIENCE/REFERENCE 0 7221 8720 3 £2.50

A selection of bestsellers from SPHERE

FICTION			
THE SKULL BENEATH THE SKIN	P. D. James	£1.95	☐
WINGS OF THE MORNING	David and Betty Beaty	£2.95	☐
THE RED DOVE	Derek Lambert	£1.95	☐
DOMINA	Barbara Wood	£2.50	☐
A PERFECT STRANGER	Danielle Steel	£1.75	☐
FILM & TV TIE-INS			
BY THE SWORD DIVIDED	Mollie Hardwick	£1.75	☐
THE DOCTOR WHO TECHNICAL MANUAL	Mark Harris	£2.50	☐
YELLOWBEARD	Graham Chapman and David Sherlock	£2.50	☐
THE KIDS FROM FAME 2	Lisa Todd	£1.75	☐
NON-FICTION			
MONEYWOMAN	Georgina O'Hara	£1.75	☐
THE FINAL DECADE	Christopher Lee	£2.50	☐
A QUESTION OF BALANCE	H.R.H. The Duke of Edinburgh	£1.50	☐
SUSAN'S STORY	Susan Hampshire	£1.75	☐
SECOND LIFE	Stephani Cook	£1.95	☐

All Sphere books are available at your local bookshop or newsagent, or can be ordered direct from the publisher. Just tick the titles you want and fill in the form below.

Name ____________________

Address ____________________

____

Write to Sphere Books, Cash Sales Department, P.O. Box 11, Falmouth, Cornwall TR10 9EN

Please enclose a cheque or postal order to the value of the cover price plus:

UK: 45p for the first book, 20p for the second book and 14p for each additional book ordered to a maximum charge of £1.63.

OVERSEAS: 75p for the first book and 21p per copy for each additional book.

BFPO & EIRE: 45p for the first book, 20p for the second book plus 14p per copy for the next 7 books, thereafter 8p per book.

Sphere Books reserve the right to show new retail prices on covers which may differ from those previously advertised in the text or elsewhere, and to increase postal rates in accordance with the PO.